I0827156

Colour and Dye Recipes of Ethiopia

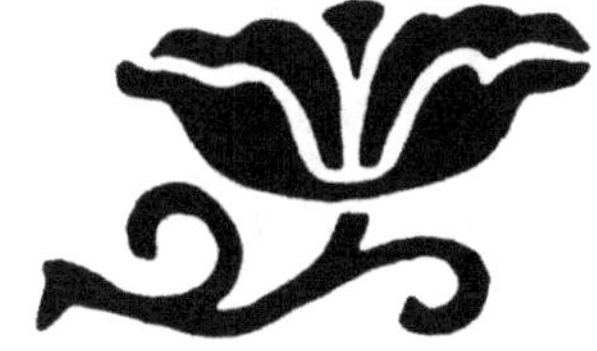

PATRICIA IRWIN TOURNERIE

Addis Ababa 1968-1973

First published privately by the author in 1986

Second edition published in 2010
by New Cross Books
for the Anglo-Ethiopian Society
(www.anglo-ethiopian.org)

ISBN 978 0 9543835 2 7

Typeset by New Cross Books, London SE14
Printed and bound in England by Lightning Source, Milton Keynes.

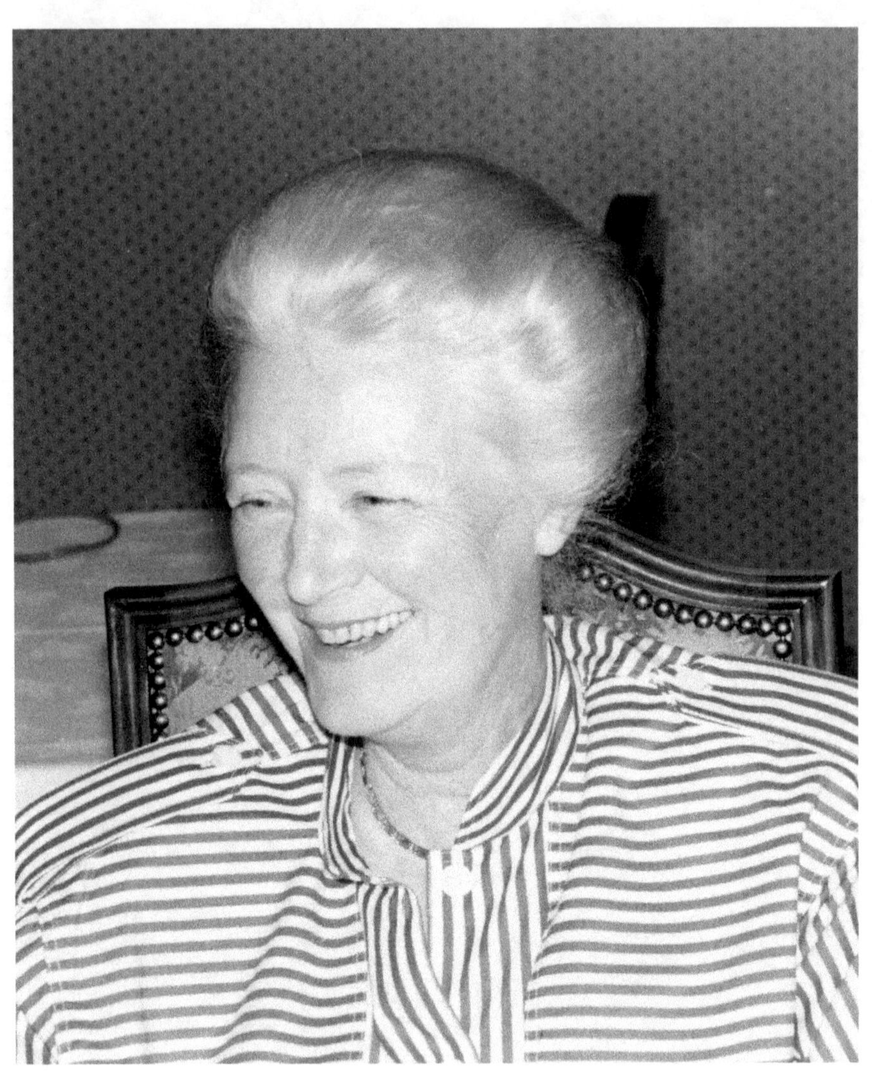

Patricia Irwin Tournerie

Patricia Tournerie – Paddy to her friends – was born in July, 1920 in Frimley, Surrey to a military family so was on the move from a very early age, living in India, Burma, Ethiopia, Eritrea and Kenya. However, it was for Ethiopia she had the deepest affection first living there in 1950/51 when her then husband, Tony Irwin, was part of the British Military Mission and again when they returned in 1955. A short period away from Ethiopia in Kenya where her marriage irretrievably broke down, Paddy returned to Addis Ababa as a single mother and worked as an art teacher at the Empress Menen School until 1963 when she returned to England for her daughter's secondary schooling.

Paddy was a gifted artist and she put that gift to good use. On return to England she enrolled at Trent Park College to obtain a Cert. Ed and taught in a primary school in Farnham, Surrey until 1968 when she returned to Addis to take up a post at the Faculty of Education at the University of Addis Ababa until 1973. It was whilst she was working with the students that she became interested in local sources for colours and inks. From that interest and through painstaking, thorough research *Colour and Dye Recipes of Ethiopia* was born.

Paddy died in June 2005. She would have been delighted and touched that the Anglo-Ethiopian Society has agreed to re-print her original privately printed volume. Even more she would have been heartened by the continued research and interest shown in the subject.

Sarah Russell (nee Irwin), 2010

Contents

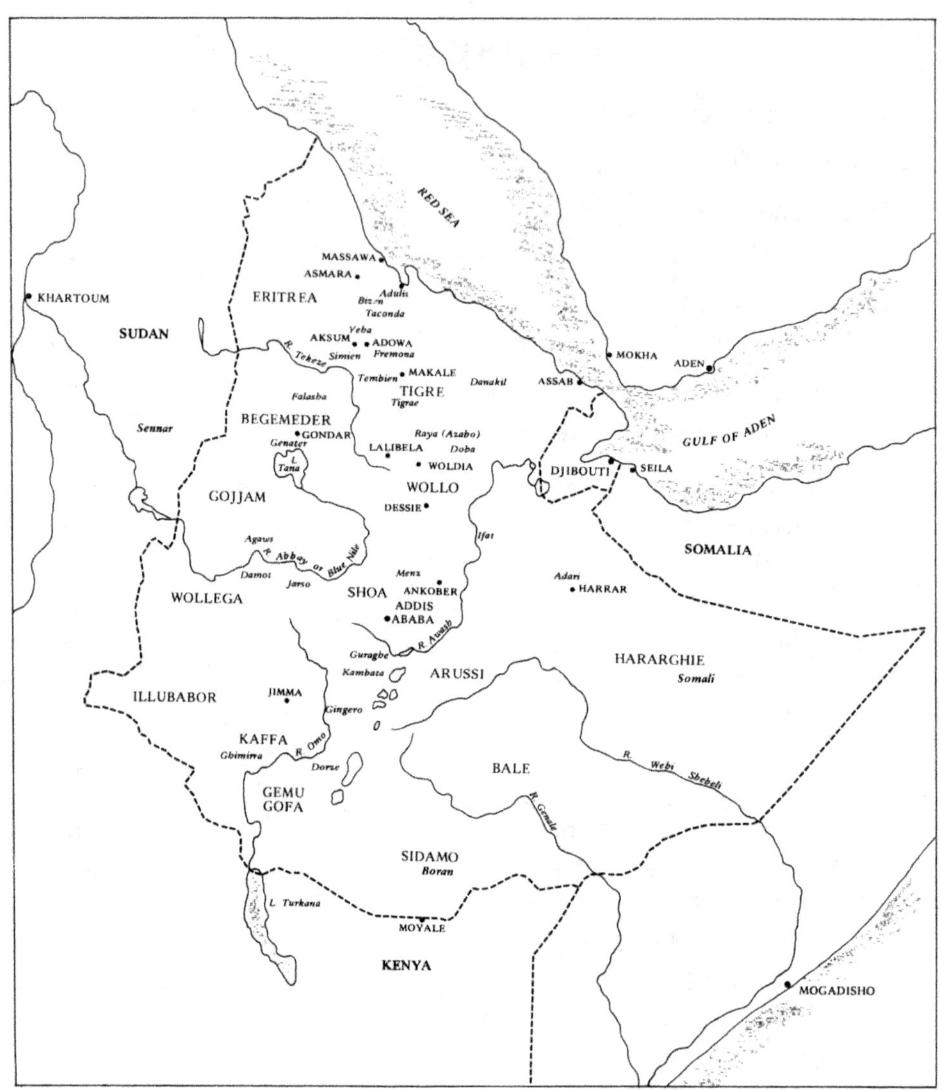
RED SEA
MASSAWA
ASMARA
KHARTOUM
ERITREA
Adulis
Taconda
SUDAN
Yeba
AKSUM
ADOWA
R. Tekeze
Simien
Fremona
MOKHA
ADEN
Tembien
MAKALE
Danakil
ASSAB
TIGRE
Tigrae
Falasha
BEGEMEDER
Sennar
GONDAR
Genater
Raya (Azabo)
LALIBELA
Doba
GULF OF ADEN
L. Tana
WOLDIA
DJIBOUTI
SEILA
GOJJAM
WOLLO
DESSIE
Ifat
Agaws
SOMALIA
R. Abbay or Blue Nile
Menz
Adari
Damot
Jarso
SHOA
ANKOBER
HARRAR
WOLLEGA
ADDIS ABABA
R. Awash
Guraghe
Kambata
ARUSSI
HARARGHIE
Somali
ILLUBABOR
JIMMA
Gingero
KAFFA
R. Omo
Ghimirra
Dorze
BALE
R. Webi Shebeli
GEMU GOFA
R. Genale
SIDAMO
Boran
L. Turkana
MOVALE
KENYA
MOGADISHO

Introductory Note

In 1968 the writer started to seek information concerning colour recipes, with the intention of providing students in the Faculty of Education at the University of Addis Ababa with a simple handbook on locally available materials for use in provincial schools.

The students themselves — the majority of whom were already practising teachers and who came from all the Provinces — were the first to be asked. Very few of them knew either dye plants or recipes, and when they did know the name of a plant often disagreed among themselves as to correct pronunciation of its name. It was largely because of these variations in pronunciation, not to mention the different languages, that it was thought unnecessary to collect names in the Amharic script. Although this may be regrettable it would not have helped the writer with the task of identifying the plants from name alone; available glossaries are written in the latin alphabet. Owing first to pressure of work and later to the troubled political and military situation, it was not possible to travel to the Provinces to investigate or verify recipes; nor in most instances to obtain specimens for identification. For this reason some of the recipes must be considered dubious and others as only of sociological interest.

In all, several hundreds of students, over a period of four or five years, were asked about dye plants. Most professed ignorance, notwithstanding their country backgrounds; but others were very helpful in bringing specimens and asking for further information from their families when they went home for vacations. These student sources are shown with no prefixes to their names.

All sources other than students have been given the prefixes *Ato* (Mr), *Wo* (*Woizero*, Mrs), or *W/t* (*Woizerit*, Miss), or their appropriate titles within the Church. It was from those connected in some way with the Church that the most interesting recipes originated. These include the recipes for red inks, most of the recipes for black inks and information concerning painting mediums and colours.

Historical notes, where appropriate, have been inserted in the recipes to give an indication of the long-standing use of a dye. At times they seem to show that a dye or its traditional use has already been forgotten.

The abbreviations following the names of plants to denote language or dialect are, unless otherwise stated, those used by H. F. Mooney in *A Glossary of Ethiopian Plant Names*. They are: (A) Amharic;

(G) Gallinya; (G-Ar) Gallinya as spoken in Arussi; (G-Sid) Gallinya as spoken in Sidamo; (G-W) Gallinya of Wollega; (G-K) Gallinya of Kaffa; (G-H) Gallinya of Harrar; (K) Kaffinya; (T) Tigrinya; (Ar) Arabic; (Som) Somali; (Ghimira) language or dialect spoken in S. Illubabor. In addition (Gurag) for Guraginya and (Kambat) for Kambatinya have been used.

Unless otherwise indicated, the local names for a plant are those given by Mooney. When new names or different pronunciations have been noted, they are shown in italics. The diacritical marks used in Part 3 of the Introduction, are those used by Professor Stephen Strelcyn.

Constant help was afforded to the writer by all those connected with the University Herbarium in Addis Ababa. In particular she was assisted by Mr Michael Gilbert, the Director, who checked the botanical names and identified specimens whenever possible, and by Dr Tewolde Berhan Gebre Egziabher.

Patricia Irwin Tournerie, 1986

Part 1:
The Dyestuffs

It seems unlikely that the art of dyeing cloth has ever been practised to any great extent in Ethiopia except perhaps in Muslim communities. Although the importation of lac, the Indian insect dye, is mentioned in the *Periplus of the Erythraean Sea*, a Greek sailing manual probably written between the late first and early third century A.D., we have no means of knowing whether it was employed for dyeing cloth or for cosmetic purposes. Most probably it was used to colour the nails and feet of the women. This practice, as in many Eastern countries, is still in vogue today, though local substitutes for lac have long been used to serve the purpose. A further reason for supposing that lac was used solely as a cosmetic is that the *Periplus*, noting the importation of 'cloaks of poor quality, dyed' and 'mallow coloured cloth', provides the first record of what was to become a continuing practice, that of importing ready-dyed materials. In support is the fact that though at this date indigo was exported from Indo Scythia (Barbaricum), at the mouth of the Indus, there are no records of its having been imported at Adulis, the ancient port of Ethiopia. Yet it was (and continued to be) bought and used in the Yemen.[1]

The dress of the early peoples of Ethiopia must remain conjectural. We have no useful descriptions until the sixteenth century, but the wearing of predominately white, or rather undyed clothing to be found in the highlands at this date might have been due originally to Greco-Roman influence acting in opposition to the traders from the East who carried the dyes. Arnaud d'Abbadie, one of the two French brothers who spent more than a decade exploring Ethiopia during the nineteenth century, is convinced that there exists a strong case for such a supposition, he finds a Greek or Roman counterpart for each article of clothing commonly worn in the highlands.[2]

A more plausible reason might have been the advent of Christianity in the fourth century. It was perhaps considered politic by the Christians to adopt, or retain, a dress which distinguished them from the Muslims who came later. In this context it is interesting to note that in present-day Yemen, whose early history was at one time linked with that of Ethiopia, we may observe fine cotton *shamma* cloth, woven in the Ethiopian manner and dyed indigo, saffron, and other colours, whereas in the Christian areas of Ethiopia the cloth remains white. Only the *tibeb* and *telat* borders carry colours.[3]

Undoubtedly there exists a strong vein of conservatism coupled with a certain lack of creative inventiveness and artistic expression among the peoples of the highlands, as all writers bear witness. Yet this alone cannot, in the first instance, account for the austerity of the garments. Far more likely is it that these traits of character ensured adherence to what was originally an edict of the church, or of church and court acting in concert. Because it was only under their auspices that fine art flourished, it is possible that dyes came in this way to be considered the province of the clergy. Certainly the only widespread dyeing of cloth was by the monks.

The question of rank played a most important part in deciding the dress of the plateau Christians. Men and women were distinguished from those higher or lower in rank than themselves by strictly defined differences in costume, visible evidence of such differences being demonstrated by the degree of colour visible in their garments. The conventions concerning dress seem to have been tightened during the sixteenth century. Baltazar Tellez, the Portuguese historian and editor of the writings of some of the later members of the sixteenth century Portuguese Mission to Ethiopia, reports: 'Within less than sixty years past, none but the Emperor, and some of his kindred and favourites, were allow'd to wear anything but breeches, and a piece of cloth to cover themselves with ...'[4] and by the nineteenth century the rules appear to have become rigid, as according to the British missionary H. A. Stern: 'The softness of the web, and the depth of the red border round the bottom of this convenient garb [the *shamma*], indicates the social position of the wearer, and this is so minutely defined, that anyone who would presume to ape his betters would, in all probability, obtain a lesson or two on dress from the imperial *giraffe*-holder. Beneath the shama the aristocrat dons his silken, damask or velvet *kamees*, but this is a privilege only granted to a few magnates.'[5]

The court and notables showed a marked taste for and dependence upon foreign products. The early Portuguese travellers found that the Emperors and courtiers were dressed in imported silks, satins, velvets, taffetas and brocades.[6] Woven cloth was paid as a form of tribute, reputedly one length for every ten woven on a loom. In addition to this local product, however, imported cloth in great quantities must have been paid over. Francisco Alvarez, who came to Ethiopia in 1520 as chaplain to the Embassy of Rodrigo de Lima of Portugal, observed that the tribute to King Lebna Dengel arriving from Tigre and Bahrmedr included silks and stuffs the richness of which amazed him. Alvarez writes: 'And since every year the silks and brocades increase in number, both those that are paid to him and those that he buys, and so many are not spent, neither can they carry them all on a journey, every year they send some to be put in pits in the earth that are designed for that purpose ... As to the silks and brocades, Pero de Covilnam said that they often took them out to give them to churches and monasteries, as was done three years before our arrival, when the Prester sent large offerings to Jerusalem of the silks and brocades from the caves, because of the multitude he possessed.'[7]

Considerable stocks were either imported or still existed in the country some twenty years later. Despite the ravages of Mohammed Gran, the Muslim invader, Castanhoso, a Portuguese officer in the party of Christopher da Gama who landed at Massawa in 1541, describes Queen Sebla Wongal on her descent from her 'strong hill' after two years of virtual isolation as being 'all covered to the ground with silk, with a large flowing cloak.'[8] Even more surprising in what was a war-torn country, is it to read that King Galawdewos, when he met this party of Portuguese, was able to present them all with 'silken tunics and breeches, for such is the country wear ...'[9] Three hundred years later Walter Plowden, the sometime English consular agent, noted that 'a silk shirt bestowed by the Ras is the distinction of the Abyssinian nobility, and this belongs only to the military class. Once bestowed, the honour is for life.'[10]

A light on early trade with the interior and the penetration of the foreign product is thrown by King Galawdewos who begged the Portuguese not to leave, trying to persuade them to stay by tempting them with the gold they could acquire from the Caffres, the negroes of Damot, who 'trade it for inferior and coarse Indian cloths, and beads of red, blue and green earth, which they valued highly and the gold very little.'[11] Bermudez, who spent some thirty years in the country, heard of traders in the heathen Guragé area between the Awash and the Omo rivers. 'The men of the province say that white men come there to trade, but they do not know their nationality, whether they are Portuguese, Turks or others.'[12] Tellez also writes of the King of Gingiro exchanging slaves for 'rare goods, brought him by foreign merchants.'[13]

This taste for imported cloth, probably well established at this date, was to remain unchanged during the next four hundred years. It is even possible that the trade in coloured cloth, by presenting the people with the finished product, discouraged initiative and experiment with local dyes, in spite of the fact that cotton grew wild for the picking and sheep were plentiful. But indigenous cotton can hardly have been sufficient for all needs, nor indeed available in all areas. Charles Jacques Poncet, the French physician who arrived in Gondar in 1699 by way of the Sudan, tells of the King of Sennar having in the town of Girana an agent who shared with the Emperor of Ethiopia the customs duties on cotton imported from the Sudan.[14] Sir James Bruce, the Scottish explorer, writes seventy years later, of both spun and unspun cotton being imported from the Yemen to Massawa.'[15] Woven cotton played an important part in the economy; it served not only for apparel but was also used as a form of currency for trade purposes, as well as for tribute. Tellez writes of 'fine *Ethiopian* cloth'[16] and later travellers found quite considerable quantities of local cotton. In the nineteenth century, the Protestant missionary, the Reverend J. L. Krapf, describes 'a quantity of cotton in Kaffa':[17] Major Cornwallis Harris, the English envoy to Shoa, speaks of the 'raw cotton which is as cheap as it is excellent and abundant':[18] Plowden, to the north, reports 'the dress of all classes is formed of the cotton cloth of the country, spun by women and woven on the handloom, principally by Mussulmans in the Teegray district',[19] and also of the Gallas south of the Nile having 'thick cotton cloth, spun by the women'.[20] The explorer Sir Richard Burton observed that woven cotton was even exported from Harrar.[21]

Whether or not the cotton was imported or local one thing is certain, and that is that weaving was an early and much-practised art. According to Abbadie, at one time it attained a very high level of excellence in Gondar. He writes of the old men who still remembered a *shamma* ornamented all over with designs of different colours woven into the cloth, but this was for the use of the Emperor and the highest dignitaries alone. The fashion, falling into disuse in the Christian provinces, was later only worn by the 'Ilmormas' to the south, neighbours of the Kingdom of Kaffa, who apparently continued to make this traditional garment.[22] Possibly this was an early example of the *dereb*, the wide multi-coloured border of complicated design whose use was confined at the turn of the century to the Imperial family and the nobility. Abbadie, Stern and Plowden are the first to describe the ceremonial *shamma* as having a silk border, or *tibeb*, woven in diamond shapes or checks. These silks, probably imported, were also used for embroidery on the women's dresses or tunics and the legs of their riding trousers. Stern remarks of the women that 'fond as they are of embroidery, high and low alike depend on their male friends for every stitch in their dress'.[23]

Apart from the white cotton, natural black woollen cloth was also woven. Harris, writing of Menz, tells us that 'the only tribute paid, therefore, is in *sekdat*, a black woollen cloth woven of the raven fleece of the native sheep of the country, and invariably employed in the manufacture of the royal tents. This fabric also furnishes a costume indispensable in so rigorous a climate ... their black weed-like habilments, distinguish them amongst all other subjects of Shoa ...'[24] There exists a sixteenth-century reference in the notes of Alvarez' secretary, which may refer to this same cloth: '... there is a very cold country where they wear burel.' Burel was a coarse woollen cloth often used for the habits of monks and friars.[25] However there was no widespread weaving of wool other than this black cloth, though Marcel Cohen, the great linguist, mentions that it was not altogether unknown.[26] James Theodore Bent, who travelled in the north with his wife shortly before the turn of this century, describes 'caps and coverlets of coarse black sheep's hair', worn at Taconda,[27] and the sportsman and traveller Powell-Cotton, writing a year or two later, finds the caps of goat hair worn in the Simien as 'woven in patterns similar to those of the Baltis in Little Tibet use for their sleeping mats'.[28] From his photographs, it appears that two natural colours were used. Some undyed sheeps' wool capes with hoods attached are today woven in Wollo. Ethiopia has many light-coloured sheep; it is possible that had it been the practice to weave wool, we should have heard more of dyes, wool in general taking colour better than cotton.

Dyed wool was, however, used in the manufacture of carpets. This small industry is described by George Annesley, Viscount Valentia, who in 1804 sent his secretary Henry Salt and three other Englishmen on an unofficial Mission into the interior from Massawa. Salt writes: 'There is a manufactory of small carpets carried on in the province of Samen, some of which were shown to me at Adowa, and they really were much superior to what might have been expected, as the production of Abyssinian workmanship'.[29] Valentia, who collated the information of his party, describes them as: 'Coarse carpets that are made in Samen, and at Gondar, from the wool and hair of the sheep and goats, which are dyed red and light blue: the former from a tree called Haddie [unidentified], the latter from a plant resembling *Indigofera*, ... they have no dark blue'.[30]

It is during the sixteenth century that we first read of the monks being garbed in yellow, and this must almost certainly indicate the use of indigenous dyes, the more so since there are no reports of yellow dyes being imported. Alvarez, writing of the monks at Bisan, notes: 'The clothes that they wear are old yellow cotton stuffs',[31] and again later, 'the monks are decent in their habits, which are full, and reach to the ground. Some wear yellow habits of coarse cotton stuff, others habits of tanned goat skins like wide breeches, also yellow'.[32] Tellez describes a leather upper garment. 'Many of those who profess the Eremetical Life wear Skins hollow'd about the Neck, and dy'd yellow, or else cloth of the same colour.'[33] At the turn of the seventeenth century, Poncet writes of the monks: '... out of church their habit is almost secular; they are only distinguished by a yellow or blue [*violette*] calot [cap = calotte] which they wear upon their heads. Those different colours distinguish their orders'.[34] After visiting the Church of St Heleni at Aksum, he wrote: 'The religious of that church are habited in yellow skins, and wear a little cap of the same colour and skin'. The monks of Bizan were described in like manner.[35] Bruce writes of a different costume for summer and winter. 'They are distinguished only from the laity by a yellow cowl, or cap, on their head. The cloth they wear round them is likewise yellow, but in winter they wear skins dyed of the same colour'.[36]

Of the nineteenth-century writers, where Harris presumes that the 'yellow dresses' are 'the badge of poverty'[37] and Charles Johnston, the British naval surgeon and traveller, describes 'the usual soft yellow leathern cape of his order',[38] an amusing though perhaps libellous sidelight is thrown on the wearers of the yellow robe by Salt. 'Above Temben, to the westward of Waldubba [is an area] celebrated for the resort of numerous pilgrims, professedly devout, who clothe themselves in a yellow dress, with a cord round their waists, and pass their time in religious and secluded retirement; but the satirical vein of pleasantry which distinguishes the Tigrians ascribes to them more unseemly motives, and scandal does not hesitate to say that Love, not of the purest kind, presides over their retreat'.[39] It is Salt, too, who describes a dress other than yellow: 'Near Genater we passed two priests dressed in light scarlet garments'. This might be a reference to the use of *Osyris abyssinica*, which is generally a red dye for leather, but may sometimes be used for monks' robes, and is said to fade to a lighter colour.[40] *Debteras*, the deacons of the Ethiopian church, are more free to wear the colour of their choice than are the monks or priests.

The colour yellow was also used for mourning. Harris tells us that '... during the period of mourning, which extends to one year, black or yellow garments, or the ordinary apparel steeped in mire must be worn as weeds ...'[41] The principal yellow dyes in use were *Carthamus tinctorius*, mentioned by Bruce, Harris and Augustus Blandy Wylde, sometime British Vice-Consul for the Red Sea; *Terminalia brownii* mentioned under the name of *weiba* by Th. Lefebvre, the French naval officer who led a scientific mission in 1839-1843, and by the French traveller Capitaine Alexandre Girard; and *Solanum spp.* mentioned by Bruce, and also by Antonio Cecchi, the Italian geographer.[42] Harris notes that '*Berberis tinctoria* of the forests yields a good yellow dye for mourning apparel',[43] but this must be a mistaken identification. This plant is not native to Ethiopia.

Red was the Muslim colour, and as Johnston puts it: '... for red head dresses of cotton cloth, and long red gowns, are invariably the "outward and visible" sign of the profession of Islamism among women of Efat, and other Mahamedan provinces, as the blue martab [neck cord] is of the Christian population'.[44] Harris describes the women whom he met on ascending the highlands below Ankober as '... merry groups of hooded women, decked in scarlet and crimson ... and a moslem dame and her three daughters ... all clothed alike in scarlet habilments ...'[45] And again: 'The women [dressed in an] ample smock of red cloth, dyed purple with accumulated lard, and the nun-like hood of the same material, buttoned close under the chin.'[46] It would seem probable that the richer Muslims favoured either imported ready-dyed cloth or imported dyes. Bruce, writing of imports to Massawa, tells us that: 'The goods imported from the Arabian side are blue cotton, Surat cloths, and cochineal ditto, called Kermis:'[47] and that four 'peeks' of the red cloth were worth the price of a goat.

Writing later of what is now Hararghie province, Burton describes the 'principal wants of the country which we have traversed are coarse cotton cloth, Surat tobacco, beads, and indigo-dyed stuffs for the women's coifs'.[48] The women of the desert and semi-desert plains must always have worn indigo or black head cloths. The Portuguese described these cloths worn by the women of the 'Moors', and the same form of black head-dress is still worn today. Such cloths may always have been imported; the desert people have ever had easy access to the coastal trading vessels. On reaching Harrar, Burton notes among the imports: 'American sheeting, and other cottons, white and dyed, muslins, red shawls, silks ...'[49] He makes no mention of dyes being imported, but two British army officers, Major Hunter and Lieutenant Fullerton, writing some thirty years later, note among imports to the town: 'Dyeing materials (crimson and indigo) imported from Bombay, for local use.' Also imported at this date were cotton twist (dyed red and yellow) from England, and cotton (coloured and dyed) imported from America, England, France and India.[50]

If it were indeed the Christian faith that dictated the wearing of undyed apparal in the highlands, it is perhaps curious that King Sahle Selassie enquired of Johnston in 1840 about scarlet and indigo and dyestuffs in general. Johnston recounts, in reference to indigo, 'I was able to promise him that I would undertake to cultivate and make [it] serviceable to his people by teaching them to manufacture the dye'.[51] Both at that time and subsequently clothes, especially the *shamma*, were decorated by coloured stripes or by borders which denoted rank according to their width. The King in his enquiry about dyes may have had only the nobility in mind: Harris observes that 'In the despotic Kingdom of Shoa, the sovereign can alone purchase coloured cloth or choice goods.'[52] No doubt this was something of an exaggeration, as Plowden tells us that the *shamma* had a red border 'if the owner be a gentleman'.[53]

C. T. Beke, the scholar and traveller attached to Harris' Mission, is categoric about the lack of locally dyed cotton thread for the borders of the *shamma*. 'The white cotton cloth is the produce of this country; but they cannot dye the red, blue and yellow with which it is ornamented, and which therefore, they import from Arabia, and work in.'[54] He was mistaken about the lack of yellow dye, for the monks used it constantly. Bruce provides an interesting example of the use of blue and yellow woven together. 'They are great novices however in dyeing; the plant called suf [*Carthamus tintorius*] produces the only colour they have, which is yellow. In order to obtain a blue, to weave as a border to their cotton cloths, they unravel the blue threads of the Marowt, or blue cloth of Surat, and then weave them again with the thread which they have died with the suf.'[55] It is not clear whether the result was a green border of woven blue and yellow threads on a white ground, or whether a blue border was woven in a yellow cloth.

In this connection it is interesting to note that a 'home-made' border sometimes seen today is of mingled blue and yellow thread made by 'embroidery', rather than by weaving. The border is created by sewing small running stitches of the two colours back and forth over the required width, producing a speckled band in which the yellow usually predominates. Elderly gentlewomen used to make this border for use on occasions when the ceremonial *dereb* or the *jano* were not used.

Thus it became the common practice to weave the borders, or *telat*, of the *shamma* from unravelled imported dyed cloth. The traveller Mansfield Parkyns is explicit in his description. 'The red border of the "quarry" is made out of a sort of coarse red cotton fabric, which is imported from India. They call the red stuff "indikky", and prefer it to red thread, though the latter be of superior quality'.[56] Johnston describes seeing the women in the market selling '... the red and blue threads of unwoven cloth ...',[57] but speaks of this cloth as though it were a woollen fabric. 'A narrow border of the blue and red woollen stuff, called shumlah, woven into the cloth, is the only ornament, and these coloured stripes will be sometimes repeated at the distance of a foot from each other through the whole length of cloth.'[58] This description is reminiscent of the present day 'Dorzé cloth'. Harris also speaks of these stripes, describing 'a great Christian Governor' at Alio Amba, 'his portly figure is completely shrouded in the folds of a cotton robe, bedecked from end to end with broad crimson stripes.'[59] Much later, at the turn of the century, A. B. Wylde describes the *shammas* as being '... all homemade with the exception of the thread for the red part, which is made of English Turkey red twist.'[60] Also at this date Bent, in Asmara, writes of the *shamma*: 'It is always white ... with a red stripe, added, so say the Christian legends of Abyssinia, to represent the blood which flowed from the body of Christ.'[61] It is likely that Bent had been speaking with a member of the clergy who had, as is quite usual, used the term 'blood of Christ' to describe the colour red. There is no evidence that the red stripe was considered particularly holy; blue borders were also in vogue in certain periods and, above all, the rulers apparently did not consider this type of decoration essential for their own wear. Dr. Richard Pankhurst writes that in the nineteenth century blue was worn by the Azabo, Raya and Gafra, and in Dobo and Wollo; red in Ifat, Jarso and Guragé.[62]

Of the non-Christian Galla peoples at Goodree near the Fincha river, Plowden writes that their thick handwoven cotton cloth was 'ornamented by pieces of blue Indian stuff, cut in various shapes and inserted; red is no favourite colour among them, strange to say'.[63] This does not sound as though the blue were woven into the cloth but rather as if it were a type of crude appliqué work. 'The women wear a skirt studiously ornamented, and fringed with the blue cloth, and by way of a petticoat, a hide, by butter and assiduity rendered as soft as velvet.'[64] According to Harris, who reports hearsay of the people of the Kingdom of Susa, the dress worn was made entirely of striped material. 'The costume of the male portion of the population consists of a robe of striped red and blue cotton in alternate bands, with tight trousers and loose kilt of the same colours and material... The females ... wear red and blue striped trousers ... with a loose shift and a robe, also party-coloured.'[65]

Sufficient evidence has been produced to show that there was very little dyeing of cotton cloth by the highland Christians, with the exception of the yellow robes dyed by the monks of certain monasteries. An unsubstantiated statement however, is provided by Wylde, who reports: 'Ras Woly told me Yejju was entirely self-supporting ... that they grew their own cotton, dyed it and manufactured it.'[66]

Widespread use was made of local dyes during the Italian invasion and occupation of the nineteen-thirties, notably of *Terminalia brownii* and *Acacia spp.*, into which all white clothes were dipped for purposes of camouflage. Woollen cloth, apart from the heavy black stuff of Menz, was very rare, though the traveller Capitaine Alexandre Girard writes that leather, wool and hemp were locally dyed.[67]

Muslims dyed cotton both red and indigo with imported dyes in Harrar, probably taking advantage also of local red and brown dyes. *Wars* (*Flemingia grahamiana*), was cultivated and exported from Harrar to the Yemen. Burton noted a 'kind of acacia, here called Galol [*Acacia bussei*]. Its bark dyes cloth a dull red ...' [68] and *Acacia seyal*, 'the tall "Wadi" affords a gum useful to clothdyers.' [69] He also observed that the women of Harrar wore 'a long wide cotton shirt, ... indigo-dyed or chocolate-coloured, and ornamented with a triangle of scarlet before and behind ...' [70] It is likely that the 'chocolate coloured' cloth was home-dyed as there are no references to imports other than crimson and indigo.

Though dyes were so seldom employed for clothing, they were in use for artifacts, especially basketry and leather work. Baskets, particularly the *mesob*, which is a combined lidded communal eating dish and portable table, were beautifully woven and certainly dyed with local colours in the past, many recipes surviving to this day. Tellez writes of these baskets as: 'Macobos, with high Lids, like Caps, the whole made of Straw or Rushes of several Colours.'[71] Although Alvarez describes the travelling bread baskets as 'painted', it is almost certain that they were actually woven or dyed reeds.[72] William Winstanley, who visited Ethiopia in the late nineteenth century, writes: 'The bread is carried in very neat wicker baskets of many coloured patterns, and is constructed exactly to hold it ...'[73] and George Montandon, the Swiss explorer, visiting the country about thirty years later, shows photographs of examples of basketry, observing that the colours of the Jimma baskets are more subdued than those of Shoa.[74] Bent gives an attractive description of the different kinds of baskets in the north of the country: 'Their great baskets for holding the cakes of bread are intricately woven with different coloured grasses and are exceedingly pretty. They also make their dish covers in the same way, and their red pepper pots, with a hole in the middle, so that the servants may wear them on their wrists as they carry their dishes to the table.'[75] He shows a photograph of this circular 'pepper pot', which is most ingeniously worked. In the opinion of Doctor P. Mérab, physician to the Emperor Menelik, the best baskets were woven by the Tigreans and the Gojjamis; but he asserts that the colours used all came from Germany.[76] There would have been considerable imports of dyes by this date, a fact of which he was aware; nonetheless his statement must be in error. Marcel Cohen describes many different types of baskets: those woven of natural straw; those rather more elaborate, with darker straw to make patterns; and baskets formed of dyed straws of different colours. He observes that ladies of good society amused themselves by weaving the latter.[77] In recent years the basket work of Harrar has been renowned for its elaborate patterns and beautiful colours.

In addition to baskets, other articles were also dyed in Harrar. Burton describes porringers '... shaped with a hatchet, finished with a knife, stained red, and brightly polished.' He adds that '... the gourd is a conspicuous article in Harrar households; smoked inside and fitted with a cover of the same material, it serves as a cup, bottle, pipe, and water-skin.'[78] Today the Qottu Galla round Harrar colour their gourds by filling them with dye left inside for two or three weeks to impregnate them, after which they are smoked with olive wood to scent them sweetly for carrying water. Imported colours have also, from time to time, been used on wood. Burton, when south of Koralay on his way to Harrar, describes his meeting with an emissary, '... in token of his sincerity, he forwarded his baton, a knobstick about two feet long, painted in rings of Cutch colours, red, black, and yellow alternately, and garnished on the summit with a ball of similar material.'[79] Some articles, including a spinning wheel, some plates and a box from Harrar, coloured in this manner, are to be found in the Ethnological Museum in Addis Ababa.

There is little historical information concerning the exterior or interior decoration of houses. Wylde gives us a description of the interior of the better class of house. 'The rafters and ties are generally most neatly worked and generally covered with different coloured cloth or painted. The walls will also be plastered with clay and finely chopped straw, and perhaps whitewashed or coloured a chrome yellow.'[80] Different local soils, and occasionally vegetable dyes, are used today to colour-wash walls in some country districts. It is only when travelling comparatively far afield that one may discover anything more enterprising. Montandon came upon decorated huts in Gimirra, finding exteriors with brightly-coloured chevron patterns. Vermilion and orange-yellow earths together with charcoal and white ash were used, the designs almost invariably incorporating the triangle, which he describes as being virtually the only motif.[81] The interiors of some of the houses in Tigré are quite elaborately painted. Ash is used in Wollega for simple decoration; it is first powdered and then stamped onto damp *chika* (mud plaster) with the base of a bottle.

In the muslim city of Harrar, where some decoration might have been expected, we learn from Burton that it was forbidden. He describes the Sultan's State Hall as '... a mere shed, ... a windowless barn of rough stone and reddish clay, with no other insignia but a thin coat of whitewash over the door. This is the royal and wazirial distinction at Harrar, where no lesser man may stucco the walls of his house.'[82] This is no longer the case today; whitewash is much in evidence in the old city.

In a country where cattle abound and where the mule or horse has always been an important means of transport, it is not surprising to find that the craft of tanning has long been practised in Ethiopia. Some hides are very roughly cured but others, as Plowden puts it, may be 'as soft as velvet'. Apart from the yellow-coloured skin garments worn by the monks, skins were cured to wear as *lamd* - the small hides complete with their legs, originally the distinction of proven warriors - and as dresses by the Galla women. They were also used as carpets, as bed coverings, and sacks. Baskets were rimmed with leather and carrying baskets were covered with it. Leather was also used for harness, for cartridge belts, and to cover various house hold or travelling articles, including books. On the caravan routes large water containers were made of leather; hides were also used for shields and for *ankelba*, which are skins in which to carry babies.

A variety of tree barks, together with the juices and fruits of plants was used in tanning, the colour obtained being perhaps incidental to the main purpose. Bruce writes: 'They tan hides to great perfection in Tigre, but for one purpose only. [?] They take off the hair with the juice of two plants, the merjombey, a species of solanum, and the juice of the kol-quall [*Euphorbia sp.*], both of these are produced in abundance in the province.'[83] Antonio Cecchi, with the Italian geographic mission, also writes of *Solanum sp.* and found it used for the yellow garments of pilgrims and monks.[84] But the tanning and colouring agent in greatest use for yellow was probably *Terminalia brownii*, mentioned by Théophile Lefebvre (one of the French party who visited Ethiopia in 1840),[85] although Girard in 1873, writes of it as being used only for dyeing cotton.[86]

Red leather was common. Johnston writes: 'It is very probable that the celebrated Morocco leather derives its bright red colour from the bark employed in tanning being obtained either from the Kantuffa

[*Pterolobium stellatum*] or the Adal tree [*Acacia sp.*], for both these trees give a very red colour to the skins that are prepared with their bark', and he describes the 'strong red infusion' which resulted after a few days from the pounded bark of *Pterolobium stellatum* to which water was added.[87] Harris too, notes the use of this bark.[88] Red leather was also obtained from the use of *kerat*. Confusingly *Osyris abyssinica*, *Acacia etbaica* and other *Acacia spp.* are all known by this name, and all are used for tanning. Lefebvre, Girard and Cecchi speak of *kerat* by name, giving no other identification. Lefebvre and Cecchi mention the dried flowers of another plant, possibly *Cassia singueana*, being mixed with kerat. The traveller Theodor von Heuglin specifically notes *Osyris compressa* (≡ abyssinica) by name, and says that it was used for a red dye; the botanist Cufodontis affirms that it is employed for tanning purposes. Mérab, on the other hand, identifies *kerat* only with *Acacia etbaica* and the resultant dye as orange. Like Cecchi, who speaks of ochres from the Simien, he agrees that these natural earth colours were also agents for colouring leather. Lefebvre notes lemon juice as the only mordant, though Girard also found alum in use.[89]

Apart from locally tanned and dyed leathers, there was also a trade in the imported product. At the beginning of the nineteenth century Salt saw '... different coloured skins from Egypt ...' for sale at Adowa market.[90] Nearly a hundred years later Bent writes, also of Adowa, '... when new, the leather saddle-cloths and fittings for the head-gear look remarkably vulgar, green, red and yellow leather being cut out in intricate patterns - a green cross, or house, or a lion on a red ground, and so forth. When old and tarnished they look remarkably well ... Then again the leather worker makes cases of stamped leather for *tedge* horns, for carrying knives and house hold implements in and for putting charms and books in.'[91] Much, if not all, of this coloured leather was probably imported.

Powell-Cotton writing of Addis Ababa, describes a similar product: 'The raw-hide market ... the vendors of imported dressed and dyed leather, coloured to bright red and greens for decoration of saddles, bridles and cartridge-belts.'[92] Mérab, writing of the popularity of coloured leather, maintains that it was imported from Europe, but fails to specify from which country.[93] Imported morocco leather was known as *yebahar arab*, 'Arabian sea'; a red dye associated with the red leather came to be known by the same name. Today it is imported as a crimson dyestuff in powder form; the hands of old men engaged on covering or repairing volumes become deeply stained by it. Some tooling is practised with hot irons, intricate designs occasionally being made on the covers of religious works. The natural leather is decorated only with patterns of another colour in Jimma, where hides are sometimes ornamented with geometric designs and stylized figures. Ox blood, the medium used for this purpose, gives a deep red-brown colour to natural calfskin.

Dyes have always been employed as cosmetics. Apart from the very early instance of imported lac, which as noted at the beginning of this section, was probably intended for this purpose, black oxide of antimony has for long been imported. Known in most countries as *kohl*, in Ethiopia it is pronounced *kuhl*, or *kulhi* in the north. Poncet notes 'black to blacken, which they call kool' as having a ready sale in Sennar and Ethiopia,[94] and all writers have observed well-to-do ladies affecting this eye cosmetic. According to Mérab it is not true antimony but lead sulphide, imported both as salve and cosmetic.[95]

Tattooing has known a considerable vogue in all parts of the country. Sometimes eyebrows were shaved and false eyebrows tattooed on; sometimes false eyebrows were painted, possibly with an ink. Gums were treated so as to darken them and the Christian cross often tattooed on the forehead, a practice said to have originated from an edict of King Zar'a Yakob (1434-1468). Very popular were tattooed 'necklaces'; the back of the hand and the base of the calf were also decorated, and sometimes other parts of the body also. Burton observes of the Muslim ladies of Harrar that '... stars are tattooed upon the bosom, the eyebrows are lengthened with dyes, the eyes fringed with kohl and the hands and feet stained with henna.'[96] Soot was the principal ingredient of the tattooing mixture, for preference soot taken from the underside of a metal vessel; it was often mixed with oil and sometimes, according to Mérab, with roasted and powdered *Datura stramonium* or other herbs.[97] Tattooing needles were made of bone or of hard wood, frequently the thorns of *Carissa edulis* were used.

After lac ceased to be popular several different plants made dye for palms of hands and the feet. *Rumex abyssinicus* which was used chiefly to tint, flavour and above all to preserve butter - was observed by Bruce as being employed as a cosmetic by the Agaws at Geesh. 'Moc-moca [*Rumex abyssinicus*], bruised, [is used] to preserve butter. Brides paint their feet likewise from the ancle downwards, as also their nails and palms of their hands, with this drug.'[98] It is also used to colour grass for basketry and for dyeing the robes of monks.

The plants in most common use in the highlands as hand and foot cosmetics are known as *insosilla* and *gurshered*. Probably the former is *Impatiens tinctoria*, though both Harris and Mérab disagree. Harris notes: 'Some *Balsamineae* grow in shady places: one of them, Impatiens grandis (*Girshid*), has a tuberous root, with the juice of which the women paint their palms and faces red.'[99] Mérab writes that in Shoa and the Galla lands, the juice from the '... root of *guerchet* (*Impatiens tinctoria*), the same plant that is used to dye the hands red.[100] Mérab observes that another plant called '*insossilla*' is used for the same purpose,[101] which Plowden also noted as a dye for 'fingers and toes'.[102] In fact there seem to be two different plants known by this name, though the terms have often been regarded as synonymous. It is possible that *gurshered* is *Impatiens rothii* or a *Conyza sp.* Von

Heuglin identifies *Impatiens tinctoria* with *elamie*,[103] under which name it is often known in the north. In the north also we find *elam* as a name for henna (*Lawsonia inermis*), by which name, incidentally, it is also known in the Boran language of Northern Kenya. Mérab writes that he never saw henna applied in Shoa except by the wives of Arabs and Somalis, and that it was not widely used in other parts of Ethiopia,[104] though Burton knew it in Harrar. It is however familiar to the northern Muslims, who also prepare a plant they call wild or 'fox' *elam* for the same purpose. This is probably *Indigofera arrecta*.

Black hair rinses are concocted from different plants, none of which are mentioned by writers on Ethiopia. In the far west of the country, and in the south, elaborate decorative head paints are often prepared from different coloured earths and clays.

Today in Addis Ababa where well-known brands of cosmetics are available, they are usually applied with greater discretion by Ethiopian women than by many of their sisters in the international community, and not infrequently frowned upon altogether by the older generation. This contrasts with the highly unflattering picture that Harris gives of the ladies of his day. 'Having first eradicated the eyebrow, the Amhara damsel paints a deep narrow curved line in their room with a strong permanent blue dye: thus imparting a look of vacancy and foolishness, which in the high-born dame is heightened by plastering the cheeks to the very eyes with a pigment of red ochre and fat ... the lady of rank completes her toilet by dyeing her hands and feet red with the bulb called *ensosela*, ...'.[105]

REFERENCES AND NOTES PART 1

1 Schoff, W. H. *The Periplus of the Erythraean Sea*, 1912. The lac insect is *Tachardia lacca*: it yields both dye and resin. Mallow-coloured cloth was cotton cloth dyed purple with *Hibiscus sp*.

2 Abbadie, Arnaud d' *Douze ans dans la Haute Ethiopie*, 1868, Note 1.

3 The *tibeb* is a patterned multi-coloured woven border, and the *telat* is a plain border of a single colour. The *shamma*, meaning literally 'woven cloth', is a wide stole or toga, worn as an over-garment by both men and women.

4 Tellez, B. *Travels of the Jesuits in Ethiopia*, 1710, p. 40.

5 Stern, H. A. *Wanderings among the Falashas in Abyssinia*, 1862, p. 311. The *kamees* is a kind of shirt worn hanging outside the trousers.

6 Almeida, M. de *Some Records of Ethiopia, 1580-1645*, Beckingham, G. F. & Huntingford, G. W. B., (Hakluyt Society) 1954, and others.

7 Alvarez, F. *The Prester John of the Indies*, Beckingham, G. F. & Hunting ford, G. W. B., (Hakluyt Society) 1961, pp. 447-8.

8 Castanhoso, M. de *The Portuguese Expedition to Abyssinia 1541-3*, R. S. Whiteway (Hakluyt Society) 1892, p. 18.

9 ibid., p. 74.

10 Hotten, J. C. *Abyssinia and its People*, 1868, p. 147.

11 Castanhoso, op. cit., p. 101.

12 ibid., p. 232.

13 Tellez, op. cit., p. 200.

14 Poncet, C. J. *A voyage to Aethiopia made in the year 1698, 1699 and 1700* (Hakluyt Society) 1949, p. 113.

15 Bruce, Sir J. *Travels to discover the Source of the Nile*, 1790, V, p. 250.
16 Tellez, op. cit., p. 40.
17 Krapf, J. L. *Travels, researches and missionary labours in Eastern Africa*, 1860, p. 57.
18 Harris, W. C. *The Highlands of Aethiopia*, 1844, II, p. 390.
19 Plowden W. *Travels in Abyssinia*, 1868, p. 49.
20 ibid., p. 285.
21 Burton, R. F. *First Footsteps in East Africa*, 1894, II, p. 25.
22 Abbadie, A. d' op. cit., Note 1, p. 611.
23 Stern, op. cit., p. 316.
24 Harris, op. cit., II, p. 349.
25 Alvarez, op. cit., p. 514.
26 Cohen, M. *Documents Ethnographiques d'Abyssinie*, p. 11.
27 Bent, J. T. *The Sacred City of the Ethiopians*, 1896, p. 216.
28 Powell-Cotton, P. H. G. *A Sporting Trip through Abyssinia*, 1902, p. 360.
29 Salt, H. *A voyage to Abyssinia*, 1814, p. 426.
30 Valentia, G. *Voyages and Travels*, 1809, III, p. 162.
31 Alvarez, op. cit., p. 70.
32 ibid., p. 126.
33 Tellez, op. cit., pp. 97-8.
34 Poncet, op. cit., p. 139.
35 ibid., pp. 149, 142.
36 Bruce, op. cit., V, p. 372.
37 Harris, op. cit., II, p. 310.
38 Johnston, C. *Travels in Southern Abyssinia*, 1844, p. 404.
39 Salt, op. cit., p. 491.
40 Valentia, op. cit., III, p. 23. *Keret* (*Osyris compressa*) can be used to dye clothes, and will 'fade to yellow'. Verbal information supplied by *Memereh* Kirkos Wolde Mariam.
41 Harris, op. cit., III, p. 161.
42 See the recipes for dyes for these references.
43 Harris, op. cit., II, p. 412.
44 Johnston, op. cit., p. 165.
45 Harris, op. cit., I, p. 364.
46 ibid., I, p. 385.
47 Bruce , op. cit., V. p. 250. This cloth may have acquired its name from kermes, another insect dye, similar to cochineal.
48 Burton, op. cit., I, p. 170.
49 ibid., II, p. 25.
50 Hunter, F. M. and Fullerton, J. D., Reports on Somali Land and the Harrar Province, 1885, Appendix C.
51 Johnston, op. cit., p. 418.
52 Harris, op. cit., I, p. 376.
53 Plowden, op. cit., p. 50.
54 Beke, C. T. *Diary of a journey in Abyssinia, 1840-1843*, British Library MS, p. 171.
55 Bruce, op. cit., V. p. 317.
56 Parkyns, M. *Life in Abyssinia*, 1853, pp. 254-5.
57 Johnston, op. cit., p. 238.
58 ibid., p. 236.
59 Harris, op. cit., I, p. 386.
60 Wylde, A. B. *Modern Abyssinia*, 1901, p. 247.
61 Bent, op. cit., I, p. 39.
62 Pankhurst, R. *An Economic History of Ethiopia*, 1968, p. 383.
63 Plowden, op. cit., p. 285.
64 ibid., p. 286.
65 Harris, op. cit., III, p. 80.
66 Wylde, op. cit., p. 360.

67 Girard, A. *Souvenirs d'un voyage en Abyssinie*, 1873, p. 241. It should be noted that Mérab, P., *Impressions d'Ethiopie*, 1929, p. 403, states that hemp is absolutely unknown in Ethiopia.
68 Burton, op. cit. I, p. 144.
69 ibid., I, p. 146.
70 ibid., II, p. 17.
71 Tellez, op. cit., p. 181.
72 Alvarez, op. cit., p. 337.
73 Winstanley, W. *A visit to Ethiopia*, 1881, II, p. 85.
74 Montandon, G. *Au pays Ghimirra*, 1909-1911, 1913.
75 Bent, op. cit., p. 120.
76 Mérab, P. *Impressions d'Ethiopie*, 1927-1929, III, p. 398.
77 Cohen, op. cit., p. 22.
78 Burton, op. cit., II, p. 28.
79 ibid., I, p. 176.
80 Wylde, op. cit., pp. 228-9.
81 Montandon, op. cit., p. 182.
82 Burton, op. cit., I, p. 205.
83 Bruce, op. cit., V, p. 317.
84 Cecchi, A. *Da Zeila allafrontiere del Caffe*, 1886-7, I, p. 293.
85 Lefebvre, T. and others *Voyage en Abyssinie*, 1845, III, pp. 241-4.
86 Girard, op. cit., pp. 241-2.
87 Johnston, op. cit., p. 373.
88 Harris, op. cit., II, p. 415.
89 Lefebvre, op. cit., pp. 241-4. Cecchi, op. cit., p. 293. Girard, op. cit., pp. 241-2. Heuglin, M. Th. von *Reise nach Abessinien*, 1874, p. 328. Mérab, op. cit., III, p. 402
90 Salt, op. cit., p. 425.
91 Bent, op. cit., p. 123.
92 Powell-Cotton, op. cit., p. 110.
93 Mérab op. cit., III, p. 412.
94 Poncet, op. cit., p. 107.
95 Mérab, op. cit., III, p. 389.
96 Burton, op. cit., II, p. 17.
97 Mérab, op. cit., III, p. 388.
98 Bruce, op. cit., VI, pp. 399-400.
99 Harris, op. cit., II, pp. 4-12.
100 Mérab, op. cit., III, p. 402.
101 ibid., III, p. 390.
102 Plowden, op. cit., p. 148.
103 Heuglin, von, op. cit., p. 328.
104 Mérab, op. cit., III, p. 390.
105 Harris, op. cit., III, pp. 158-9.

Part 2:
The Artists

Except in the case of the prehistoric cave paintings found in various sites in Ethiopia, there remain no examples of the use of pigments until about the eleventh century. The early peoples were stone masons par excellence, as the architectural remains of the pre Aksumite kingdom in the north, which include a fair part of a temple at Yeha, bear witness. These peoples used bronze tools, weapons and seals, and also manufactured pottery; there is evidence of the Sabaean influence upon the culture of these times.

The Aksumite empire, which came into being sometime about 500 B.C. was again notable for architecture and for stone carving. Examples are still in existence; the remarkable stelae of Aksum mirror in stone the manner in which buildings of more than one storey were constructed, using as materials both wood and stone. In the early days of this empire, Greek influence is apparent. The great palace at Aksum may have had some coloured decorations but no trace of mosaics has been found.

The advent of Christianity in the fourth century no doubt brought about a gradual eclipse of any secular arts which may have flourished before this time. The Christian symbol appeared on coins and pottery, architectural talent was turned to the construction of 'standing' churches, a few remains of which survive to this day. By the fifth century Christianity was securely established and from that date it is probable that all creative art fell within its province. The Aksumite empire, however, gradually fell into decline. While it is possible that the emphasis placed by the Church upon the life of the spirit may have inhibited the growth of civic structures, it is owing to Christianity and the existence of a written language that Ethiopia today is heir to a literary heritage and to a remarkable and individual school of mediaeval art.

In order to understand how this Christian school of art came into being, we must briefly examine the external influences to which the early Church was subject.

The Monophysite monks who fled from the Near East after the Council of Chalcedon in A.D. 451, almost certainly included in their numbers the famous Nine Saints of Rome (actually Byzantium).

According to tradition the Nine Saints were early missionaries to Ethiopia and either founded or reorganized her monasteries. Scholars are of the opinion that they were Syrians who travelled by way of Egypt and Nubia. When Cosmas Indicopleustes visited Aksum about the year 525 A.D., he reported the existence of churches, priests and Christians in considerable numbers. Later travellers have always been 'struck by the Byzantine influence apparent in early Ethiopian art; apart from these first Monophysite monks, there was indeed a Byzantine embassy to Aksum during the reign of Emperor Justinian. It is most noticeable that perfect circles and arches in the form of segments of a circle, are to be found in the early examples of illustrated manuscripts, thus demonstrating the influence of the Eastern Church. The art of drawing perfect circles was then either forgotten, or considered unimportant by Ethiopian artists until the Gondarene period in the seventeenth century. The slight Arab influence often remarked may be attributed to the artifacts carried by Arab traders and the possible employment of Arab craftsmen; perhaps also to the fact that the first Muslims who fled from persecution in Arabia were given sanctuary in Aksum in 615 A.D. Among these Muslims was a woman, who, after her return to Mecca, is reputed to have spoken with nostalgic admiration of the beauties of the cathedral at Aksum, and of the wonders that adorned its walls.'[1]

The head of the Ethiopian Church, the *Abuna*, was appointed by the Patriarch of the Church in Alexandria. We may wonder whether, when a new *Abuna* set out for Ethiopia, he was ever accompanied by craftsmen and painters. There exist no records that the *Abunas* brought with them their own manuscripts, or other examples of Christian art, but a Coptic influence is undeniable and the later rendering of equestrian saints owes much to it. Further, the persecution of Christians in Egypt in the late tenth and early eleventh centuries caused many Copts to take refuge in Ethiopia.

By the twelfth century an Ethiopian colony was in being in Jerusalem, where land on which to create a monastery had been granted. Ethiopians were in the habit of making pilgrimages to the Holy Land, and here they must have seen, and possibly brought back with them, articles in the style of the Church of Armenia.

In the tenth century a legendary heathen queen, popularly supposed to be the Agaw - Judith - ravaged Ethiopia and completed the destruction of the declining Aksumite empire. How many Christian or pre-Christian works of art were destroyed at this time, we have no means of knowing.

As a result of the seventh century Muslim invasion of North Africa and the Near East, Ethiopia was virtually isolated until the thirteenth century. It was, however, during this period of isolation that the Zagwe kings ruled in Lasta province and that the celebrated monolithic rock churches of Lalibela were built. King Lalibela reigned from circa 1190-1225; though he is credited by popular belief with their construction, the building of these churches must have taken many decades. Here again we find a breech in the Muslim wall of encirclement; tradition has it that five hundred artisans from Egypt and Jerusalem, or from Armenia, were employed as stone masons for this task. There are numerous rock churches in northern Ethiopia. It might have been the invasion of the heathen queen that caused them to be hollowed out in such inaccessible places, often on mountain peaks; on the other hand, the practice of 'hiding' churches - or of using such sites as churches - may have a much earlier origin. It is only possible to date these churches approximately by the style of mural paintings which of course may have been the creations of a later date. It is at the time when the Zagwe dynasty was overthrown, in the second half of the thirteenth century, that Coptic monks are known to have imported many of their Church's liturgical and theological works and so infused new life into the Ethiopian ecclesiastical literature. There was much copying and translating of manuscripts: examples dating from this time are fortunately still in existence.

Influences on early Ethiopian art stemming from Persia, India, and even beyond, cannot be excluded. Aksum was an early trading centre of some importance: taking into account the fact that Ethiopians at few periods in their history have had any great reputation as craftsmen, often employing foreign artisans, Aksum could have been responsible for Eastern infusion, however minimal in extent.

The Emperors, Kings, and great Rases of Ethiopia were always patrons of Church art, known frequently to have employed painters to execute works for the churches they patronised and to have caused to be written illustrated texts for their own use. If his work was considered good, the King did not hesitate to employ a foreigner. This fact certainly resulted in the old ecclesiastical painters being directly exposed to painting techniques from other lands at a time when European art was employing similar materials: parchment, wooden panels and natural pigments such as are to be found in Ethiopia.

We read that King Dawit I (1380-1412), sent to Venice the gift of four leopards, spices, the skins of monkeys and a zebra. In 1402 the Venetian Grand Council decided to repay his gift by despatching to Ethiopia a group of skilled artisans consisting of two masons, a carpenter and a metal worker, headed by a painter.[2] King Dawit, on friendly terms with Egypt, sent to them in 1387 an embassy with twenty camel-loads of gifts. It was in his reign, too, that an Italian, Pietro Rombulo, arrived in Ethiopia, in 1407. He remained for over forty years before returning to Naples in 1450 and was at one stage employed on a mission to India and China on behalf of the Ethiopian king.

During the reign of King Zar'a Yakob (1434-1468), more foreigners arrived and settled in the country. A certain Brother Battista da Imola who visited Ethiopia in 1482 met several who had at the time been resident for some twenty-five years. There were many Italians among them, apparently so well established that they had brought with them, or sent for, 'a large and ornate organ in the Italian style'[3] which they had set up in the church of Atronsa Maryam. The sight of the instrument amazed Brother Battista as much as it amazes us to learn of it today. As Zar'a Yakob was a great reformer, building and endowing many churches, it seems more than likely that among the resident foreigners was a painter or painters engaged upon decorating not only the church with the Italian organ but also undertaking other commissions at the King's request. One such painter absorbed into a religious community, was in all probability Fere Seyon. It was not the custom for Ethiopian painters to sign their work, and as we know that a decade or two later the Venetian painter Brancaleone adopted an Ethiopian name, it seems possible that the author of the following inscription was a foreigner: 'This picture was painted during the reign of our King Zár'a Ya'qob and our abbot Yeshaq of Däga. And the painter is one poor man of the Monastery of Guben, Feré-Seyon ...'[4]

The chronicle of Ba'eda Maryam (1468-1478) mentions that the King had a painting of the Virgin and Child executed by a Frank, or foreigner. Because the Child was portrayed, in the Western manner, as on the left, or inferior, arm of the Virgin, this painting was reputed to have given offence.

Only a few years later there arrived in Ethiopia two Venetian painters whom we know by name. The first was Hieronymous Bicini, subsequently taking the name of Gregorewos, who left Venice in 1482. A Venetian named Zorzi, questioned an Ethiopian monk about him in Jerusalem in 1523. He learned that: 'Presta David [Dawit being the throne name of King Lebna Dengel (1508-1540)], gave the said Venetian Bicini an estate with castles and a city under him, and he is married and has several children and rides with 70 horses, and is a secretary of the said Presta: and he has painted for the said Presta many things ...'[5] Bicini must have died before Francesco Alvarez arrived with the Portuguese Mission in 1520, as Alvarez does not mention meeting him. He did, however, meet the second Venetian painter, Nicolao Brancaleone, known locally by the name of Markoreyos, who arrived in the country either in 1480 or 1487. Alvarez saw his work in 'a church ... which is named St George ... It is a big church, with all the walls painted with suitable pictures and very good stories, well proportioned, made by a Venetian who has been mentioned before, named Nicolao Brancaliam, and his name is on these pictures, and the people here call him Marcoreos ... (and he knew the language of the country very well): he was a much respected person, (very rich) and a great lord of a big country with vassals, although a painter ... said to be a monk before he came.'[6] From remaining examples of his work it is apparent that although a rather poor draughtsman his painting ability was adequate and that he made very good use of the materials available locally.

Several other painters may have trained under him; we find examples by another brush using the same palette. Bicini was, however, the better artist, and it is reasonable to suppose that his works had a greater influence on Ethiopian painting.

When the Portuguese Mission was about to leave Ethiopia in 1526, Alvarez reports that King Lebna Dengel requested that one of their number should remain behind. This was the painter Lazaro de Andrade, a man of considerable talent, a fine painter. He was engaged in painting work for no more than a year before the Muslim invader, Mohammed Gran, started to lay waste the country. During the period that followed churches without number were destroyed, together with much Church art. Only in recent years have manuscripts and paintings hidden from the invaders by the priests come to light; while of the murals in the large churches decorated by the Venetians there still remains no trace. In 1541 the Portuguese returned to Ethiopia, a mission led by Christopher da Gama landing at the port of Massawa. Despite initial reverses, its members joined with the forces of King Galawdewos and succeeded in routing the Muslims. In 1543 Mohammed Gran was killed.

A Jesuit mission under the leadership of Bishop Andre de Oviedo arrived in Ethiopia in 1557. The Jesuits settled at Fremona, near Adowa, where they remained for about a century before being expelled by King Fasiladas. During that time Jesuit painters and draughtsmen, together with the works of European Christian art they had imported, exerted considerable influence on Ethiopian Church art.

It is one of the peculiarities of Ethiopian art that no matter the country of origin of the external influence, techniques were adopted and styles adapted in such a way that the art forms were ultimately translated into what proves to be an unmistakably Ethiopian manner. It is also interesting to observe that artistic ability in the hands of foreigners seemed generally to be uncritically admired. Even if it were not, such gifts as they possessed were made use of to a greater or lesser extent. As late as the beginning of the nineteenth century, when Nathanial Pearce, one of Viscount Valentia's unofficial mission, asked permission to remain in Ethiopia, Henry Salt, Valentia's secretary, reports the interview with the Ras, who '... questioned him with regard to his capacity as a soldier, his ability in painting, and his knowledge of physic.'[7]

If there is very little documentary evidence to suggest that paints or painting materials were imported into Ethiopia during mediaeval times, it seems certain that the various foreign painters would have brought with them the basic tools of their trade. Communication with Christian centres in other countries was maintained, paintings from abroad undoubtedly finding their way into the highlands of Ethiopia. Notable here are the 'St Luke' icons of the fifteenth century Cretan school and certain other fifteenth- and sixteenth- century works of European origin.[8] If - as today - it was the custom from earliest times to show these venerated works only very briefly on festival occasions their influence would not have been considerable on the practising artists of the time, who may never have been afforded the opportunity to study their techniques. But imported paintings less valued were freely displayed, as was noted by James Bruce in 1790: '... sometimes for a particular church, they get a number of pictures of saints, on skins of parchment, ready-finished from Cairo, in a stile very little superior to these performances of their own. They are placed like a frize, and hung in the upper part of the wall ...'[9]

The painting pigments favoured will be discussed in Part 3, but here we may make some brief observations regarding imports. A late seventeenth-century canvas from Lalibela provides evidence of the use of imported lapis lazuli:[10] this adoption of lapis lazuli is unusual in the Ethiopian context. Indigo has frequently been identified; it appears to have been imported for many centuries. Charles Jacques Poncet, the French physician who arrived in Gondar from the Sudan in 1699, observed for sale in Sennar - which he stated to have 'good vent' in Ethiopia - articles including vermilion, sublimate which may have been the pigment known as blue sublimate, and white and yellow arsenic.'[11] This was more than one hundred years after Alvarez had observed that Ethiopian priests engaged in trade. They may either have ordered materials which they required and had been taught to use by foreign artists, or have purchased the foreign pigments primarily for medicinal and magic purposes. In the early nineteenth century it must have been a normal practice to order foreign paints. Salt, describing the work of an Ethiopian artist, tells us: 'The materials employed by this artist were of the most common description, and had been brought by a Greek from Cairo.'[12]

Nathanial Pearce, who came originally to Ethiopia as a member of Salt's party, and who stayed behind and resided in the country for some dozen years, made these observations: 'Their manner of painting is, I think, very curious: it is as follows. After plastering the wall and smoothing it with clay, they line it, when perfectly dry, with cotton cloth, which is stuck to the wall by means of a slimy substance made from cow's hide, or from the fruit of the *wanzatra*. Over this cloth they lay a coat of whitewash, made from chalk or lime-stone, first burnt, and then pounded and mixed with water, adding a little of the aforesaid substance with which the cloths are stuck to the wall. They then draw the outline of the picture with charcoal, and afterwards paint it with black paint, which they make by burning hemp seed nearly to a cinder; they then shade their painting, by strengthening or weakening their colour. They make no colours in the country, except a fine red, which they use for dyeing ivory, and this is made from a wood called *zanen*. All other paints they obtain dry from Arabia; these they grind and mix them selves, and always mix the yolks of eggs and gum-water in their paints of all colours. The paints are ground on a smooth stone, with the yolks and gum-water, and tempered with the same.'[13]

Capitaine Girard, who travelled in Ethiopia in the eighteen sixties, gives us the following description. 'A *seuli* takes six months to paint a poor picture of the type that could be finished in one day by one of the least of our daubers [rapins]. Ethiopian painters have only one method: it is 'distemper': to do it they use a glue of scraps of skin and a piece of charcoal made from olive wood. When they want to paint a picture, in a church for example, they affix a 'canvas' of local cotton on the wall, prepare it with this glue, and then, with the charcoal they make the sketch: this done, they grind their colours with white of egg in bowls, and set to work. Their brushes are made of hairs from the tail of a cat or a goat that they dip alternately, according to the tint [ton] that they wish to obtain, in vermilion, lead white, ivory black, Prussian blue, Cobalt blue, green, yellow, and a certain blue composed of lead white and powdered glass. They also use verdigris.'[14] This somewhat derogatory description of the use of parchment glue and such ancient materials as verdigris and smalte - a blue similar to that employed by the ancient Egyptians - gives us a glimpse of how much lore had been handed down and how these ancient materials were used in combination with imports.

Professor Doctor C. Keller, writing in 1904, tells us he heard that '... although today people try to procure ready-made colours from merchants ... in former times there existed a very lengthy technique for the preparation of colours, which were at times mineral, and at times of organic origin.' He adds a further point of interest, namely that: 'Red colour (cochineal), was procured from the exterior and was very costly.' Keller describes both egg white and egg yolk as being used as mediums.[15]

Major Harris, leader of the British mission who came to Ankober in the first half of the nineteenth century, and who held a low opinion of the local art, gives us this description: 'The pictorial art is still far behind the middle ages of Europe; and the appearance of the limner arranging his design with a stick of charcoal, or filling in gaudy partitions with the chewed point of a reed dabbled in the yolk of an egg, which is placed on end before him, proves sufficiently diverting'[16] This is an interesting eye-witness account of how yolk of egg was put to use.

The traveller William Winstanley, who visited Ethiopia during the second half of the nineteenth century, speaks of the 'cathedral church' of Frangar. 'I was assured by the custodian of the temple who acted as guide, that artists had at an anterior date been imported from Alexandria by previous Aboonahs to effect the evidently highly esteemed works of art which adorned the walls.'[17] Dr Mérab, physician to the Emperor Menelik at the beginning of the twentieth century, goes so far as to state that colours have always been imported from abroad, except during the time of the Portuguese, who made them locally, but without finding either imitators or those who would continue to make them. He also confirms the use of white of egg and glue of beef hides.'[18]

This view is refuted by Paolo Graziosi, writing in 1932, who tells us: 'Red was obtained from the ground bark of an abyssinian plant, the 'Kerate' [*Osyris abyssinica*], mixed with the flowers of another plant, the 'Bebeteia' [*?Cassia singueana*]. Yellow was made with the bark of 'Ueba' [*Terminalia brownii*], blue with an earth from the Simien, and green with ground glass and lead white. These colours were mixed separately with white of egg, then they were diluted with the juice of a gummy plant. These were, and still are, spread directly onto the parchment without any special preparation, while for the frescoes a cloth on which to paint was attached to the wall plastered with a mixture of gum and chalk.'[19] Dr Mérab is no doubt only repeating what had been told to him. There is some truth in his assertions; old painters of the present day whose fathers were also artists say that they, and their fathers before them - who would have been painting in Mérab's day - always preferred to use imported paints when possible. This was not, however, a rule. Many old painters still living, as well as those who have but lately died, took considerable pride in making all their own colours and jealously guarded the recipes.

If some reference to imported paints, imported pictures, and imported artists is made at all periods, throughout the centuries local materials have also been used by Ethiopian artists.

REFERENCES AND NOTES PART 2

1 Gerster, G. *Churches in Rock*, 1970, p. 62. Leroy quoting Tabari.
2 Conti Rossini, C. 'Un codice ilustrato eritreo del secolo XV'. *Africa Italiana* 1, 1927, p. 88.
3 ibid., p. 92.
4 Chojnacki, S. - 'Notes on Art in Ethiopia in the 15th and early 16th Century', *Journal of Ethiopian Studies*, Vol. VIII, No. 2, p. 38.
5 Conti Rossini, op. cit., p. 94.
6 Alvarez, op. cit., pp. 279, 313, 332.
7 Valentia, op. cit., III, p. 143.
8 Spencer, D. 'In Search of St Luke Ikons in Ethiopia', *Journal of Ethiopian Studies*, Vol. X, No. 2, pp. 75, 76, 88.
9 Bruce, op. cit., Vol. V, 12, p. 315.
10 Chojnacki, S. - 'Notes on Art in Ethiopia in the 16th Century', *Journal of Ethiopian Studies*, Vol. IX, No. 2, p. 89.

11 Poncet, op. cit., p. 107.
12 Salt, op. cit., p. 395.
13 Pearce, N. *Life and Adventures in Abyssinia, London*, 1831, reprinted by Sasor, London, 1980, Vol. I, pp. 224-45. The writer has unfortunately been unable to identify *wanzatra* and *zanen*. *Wanza* is *Cordia africana* (A).
14 Girard, op. cit., p. 178.
15 Keller, C. *Uber Maler und Malerei in Abessinien*, 1904, p. 36.
16 Harris, op. cit., p. 182.
17 Winstanley, op. cit., p. 125.
18 Mérab, op. cit., III, p. 304.
19 Graziosi, P. *Le Pitture Ethiopiche de Museo Nazionale d'antropologia ed etnologia de Firenze*, 1932, p. 5.

Part 3: The Colours

Ignorance of dye plants and of the old methods of producing dyes is now the rule rather than the exception in all countries, men and women having come to rely on chemical dyes and on the finished product. The young people of Ethiopia, particularly the educated and those living in urban areas, are not exceptions. Their grandmothers may know how to prepare dyes for basket work and how to darken their clothes for mourning purposes, but their mothers probably rely on imported chemical dyes. Some girls still colour hands and feet with henna (*Lawsonia inermis*), or *insosilla* (*Impatiens tinctoria*) for cosmetic purposes. But their grandfathers may remember how to dye leather for harnesses or how to colour mats: certainly they will remember dyeing their clothes for camouflage purposes during the Italian war and subsequent occupation. The children know little by comparison. In country districts they may help to dig the coloured clays and earths used in colour-washing and decorating their homes, play with berries and plants to make little patterns on their clothes and produce 'inks' they use in school; but they are often unable even to describe colours with any exactitude. 'Black' for example is a word meaning any dark colour, brown, dark blue or purple.

Only among the clergy does the tradition of making and using colours still live in any real sense. Many monks still dye their yellow robes with the old dyes. Most elderly priests still guard their own recipes for making very fine black inks: fewer know how to prepare red inks. Those who have received the traditional training as artists within the Church, or whose fathers were trained within its auspices, can remember some of the recipes for making colours, but for many years now have found it easier and more practical to use imported products.

Unlike lay persons, the clergy have always used a reasonably exact terminology to describe tints. A shade of red may be *däma kristos*, Blood of Christ, or *nädd*, flame; a certain green *keteliya* or *ketelima*, leaf green, from *ketel*, leaf; a rich blue is *libse mariam*, Mary's robe; and a lighter blue is known as *nil*. According to folk tradition this is said to derive from Nile blue, a term having been learned by pilgrims to

Jerusalem. In fact the word is almost certainly from *nīl* or *anil*, the Arabic word for indigo.

Traditionally colours have always been endowed with significance. According to the Ethiopian philosophers there are reputed to be four colour 'families' related to the four elements. These four 'families' are represented by the colours black, white, yellow and red. One follower of the philosophers has verbally described them as follows:

> 'Black is the colour of the element Air, and the wind is considered to be black. Black is also the colour for God the Father, because His mysteries are unfathomable. Included in the family of black, and considered to be related colours, are the blues and many of the greens. The deep tints of most colours are judged to be black.
>
> White is the colour of the element Water, and water itself is considered to be white. Drinking water makes a person white, as does prolonged bathing. The colour for Heaven is white, like the first light of dawn: white is also the colour for faith, and represents purity.
>
> The colour of the element Earth is yellow, and it is the sign of fertility as the first shoots of grain are yellow. Yellow is also the colour for God the Holy Ghost. Ripe fruit, a pregnant woman, and even a pregnant locust become yellow, and with a dying person it is because the Holy Ghost is trying to lead him to God.
>
> Red is the colour of the element Fire. If a person is struck in the face, the part becomes red as the blood wishes to escape: this is the reaction of nature when exposed to heat or fire. Red, as a symbol of blood, is the colour for God the Son. The colour brown belongs to the family of red.'[1]

It is probable that the concept of four colour families is derived from very ancient sources. For example, in ancient Egyptian art, we find red for Fire, yellow or gold for Air, Blue for Water, and green for Earth, symbolising the four elements, the four seasons, and the humours. Possibly in Ethiopia the meanings associated with the colours changed through the centuries and with the advent of Christianity. Even in more recent times there are many varying explanations given for the symbolism of the colours in the Ethiopian flag. Some say the colours represent the Trinity, with yellow for the Father, red for the Son, and green for the Holy Ghost. The same colours, though differing from Dante's symbols of the Christian virtues, except for green, which represents Hope in both cases, have been described as meaning Faith, Hope and Charity. Professor Chojnacki notes that the three colours are dominant in several old Ethiopian manuscripts, especially during the Gondarene period.[2] Colours used in the manuscripts and paintings must have been dictated by available materials as well as by the Church Canon of the

Middle Ages which gave significance to the colours of liturgical garments. This is reflected in Ethiopian paintings, where the Virgin's robe, as elsewhere in the Christian world, is generally of the classic blue, indeed lending its name in Amharic to the same blue.

The *haräg*, a vine or an entwined basket pattern frequently used as decoration in manuscripts - most often at the beginning of a chapter or surrounding a holy picture - is said to symbolize the three aspects of God. God the Father is represented by the colour *nil*, a pale azure or pale indigo blue, which belongs in the family of black. God the Son is represented by red, sometimes called *däma kristos*, Blood of Christ. God the Holy Ghost is represented by Keteliya, the leaf green which is part of the family of yellow. Other priestly painters hold that God the Father is represented by blue, that being the colour of Heaven; God the Son shown by red, being the colour of the flesh; and God the Holy Spirit by yellow, as belonging both to the Father and to the Son. Similar use of the three colours representing the three aspects of God may be noted in the custom of tying a tri-coloured cord round the neck of a child at christening.

In the days before both boys and girls with artistic ability could enter an Art School, the training of artists took place within the Church. Young children attending a church school - there were few girls - received their first drawing lessons from a priest or *debtera* (deacon) sitting in the open under a tree. They practised with both charcoal and pens made of *shimboko* (*Arundo donax*) on the cleaned shoulder-blades of animals. When work was not up to standard they were told to wash it off and start again. This tedious process was repeated times without number. Boys who then showed a real aptitude for drawing were set to copy certain traditional subjects: little landscapes with a few round huts, a priest in his cloak, stylised portraits with over-large eyes. The most promising pupils were sometimes allowed to use off-cuts of parchment - scraps from the edges of the skin - but even these were precious and never to be wasted.

The young artists had to learn how to prepare parchment, a task one may still see performed in the country districts. The animal skins are roughly cleaned and then taken to the river to be left in the water for four or five days. Sometimes they are carried down from the mountains to a lowland river, where the warmer water proves more effective. After this the wet skins are stretched well off the ground between posts, to which they are tied with leather thongs. When the damp skins have been evenly stretched they are very carefully scraped to take off all the hair. Skins are usually re-washed three or four times and often rubbed with pumice stone, both to whiten the skin and to make the surface very slightly rough, like canvas, lest the ink should 'slide' on the finished 'page'. The ordinary parchment is generally goat skin but the finest and largest is made from the skins of horses, which have fewer blemishes. Great care is taken with good parchment; it is always kept wrapped in *shash*, fine handwoven cotton muslin, to prevent dust or dirt from getting to it.

Much time was spent in learning how to make ink. The inks used on Ethiopian manuscripts are of very fine quality, and have remained completely black after centuries, defying the ravages of time and weather. There are many recipes, but most of the good inks are a combination of the classic blacks: ivory black, vine black and lamp-black mixed with various other ingredients. The making of fine inks is an extremely lengthy and laborious process, taking anything up to six months. As all inks are sun-dried the ingredients are sometimes transported to the lowlands in order to hasten the process a little in a warmer climate.

Some ink recipes contain a kind of insecticide to prevent flies from spoiling the page or disturbing the calligrapher while he works. Most contain a little resin or gum, the best being supposedly a sort of *itan* (incense, generally *Boswellia papyrifera*), which has itself to be collected from a lowland area, ground and strained, so that it becomes white and completely free from impurities. Kept in tablet form, it is melted in the sun with a little water when needed. The quantity used in the ink has to be nicely judged; if there is too much the parchment pages may later stick together in wet weather. A finished ink of good quality is completely solid, shiny black and without the slightest crack. A noted calligrapher still living, whose ink was much sought after, used to charge one Maria Theresa dollar for a small cube. When ink is required a small piece of solid ink which must be dissolved before it can be used, is broken off, the dissolving process taking two or three days. The fluid ink is kept in a *qäläm kend*, literally ink horn, a pointed horn vessel which can be stuck into the ground.

The young pupils also learned to gather flower petals and different coloured earths to make painting colours. Petals were dried and powdered, earths sieved and ground. A little resin would be added and then, for painting or writing, yolk of egg used to bind the colour. Sometimes there might also be a few drops of garlic juice added to ensure a good gloss, though here again too much might cause parchment pages to stick.

Many stories circulated among the gifted boys. There were tales of hardships to be borne in the monasteries; there were tales of the 'library' in the monastery of Waldebba, housed in huge caves in which armed guards kept watch continually over the treasures of the Church; there were tales of how good artists were honoured and plentifully supplied with food and drink by their patrons. But there were also stories about important men who had commissioned especially beautiful manuscripts or icons and who, determined that no one else should possess similar works, would cut off one or both of the artist's hands so that he could no longer work, thereafter feeding him out of charity.

A monk learns a variety of crafts. Some monasteries are entirely self-supporting, all agricultural work and animal husbandry shared between the monks. In most there is a rotation of tasks: in Bizen, for example, four years is the general term, after which a change of duties is ordered by the Abbot, often according to age and aptitude. If a scribe's eyesight starts to fail, he will be allotted a new task, that of prayer alone not usually being granted before a monk attains the age of sixty. Sometimes humble work is sought after by the pious as bestowing increased grace, though in theory all tasks rate equally.

The painting training is a long one and young aspiring clergy and monks are first assigned to mural panels. A good cotton cloth is tacked to the wall and covered with coatings of chalk and *mashega*, animal skin glue. Juice from the native olive tree (*Olea africana*), is then rubbed over this base with a cloth before the outlines are drawn in charcoal for the picture. Several years must be spent on mural work under a master before an artist is considered sufficiently proficient to start experimenting on parchment. Generally an artist trains himself in the later stages.

Before starting a painting, a priest will withdraw into solitude to meditate and to pray that his hand may be guided by God - that a vision of what he should paint may be given him. In small communities the same man may prepare his inks, copy and illustrate a manuscript, sew the pages, prepare the boards for the covers and bind them with leather. Only in an area such as Gondar, where the output of manuscripts from many pens is considerable, are there specialists. There, one man may stitch the pages, another prepare the olive- or cedar-wood covers, a third bind and emboss the leather of the finished volume.

Good calligraphers do not work on Saturdays or Sundays, and on Mondays are forbidden to continue with any important task they may be engaged upon. They have to spend the day practising, starting work again on the Tuesday.

Magic continues to play a part in the lives of many of the people. Religious texts and special prayers against evil or disease are copied on to long narrow strips of parchment which are then tightly rolled and worn as amulets. In times of sickness they may be opened and hung where the sufferer can view them. The magic scrolls are illustrated with guardian angels, saints, and various protective devices; text, pictures, and colours to be employed are prescribed in the magical and astrological book, the *Awde negist*, 'The circles of the kings'. The composition of the inks used in the scrolls may differ, various special ingredients being added as necessary. Thus a scribe who copies certain passages of the scriptures for a magic scroll will use the ink specified in the recipe for magic, but when copying the same passage for his church will use his usual ink.

Although in theory prescribed, the recipes for the inks and colours used in magic scrolls vary from man to man and from village to village, perhaps depending to some extent on available materials. The Falasha peoples, reputed to be the descendants of one of the lost tribes of Israel, have acquired a great, probably exaggerated reputation for magic. Except, however, for the introductory wording of the passage or prayer, their scrolls differ little from others.[3]

Readily available in the markets today is an imported powdered crimson dye. This colour, frequently employed for magic scrolls, is mixed with yellow to give scarlet when it is to be used for church purposes. The powdered dyestuff is a modern *nädd,* 'flame' colour, *nädd* playing a significant role in magic both on its own and as one of the *säbatu qälämat,* the 'Seven Colours'. The mixture known as the *säbatu qälämat* is used in traditional Ethiopian medicine and in magic. As a medicine it is considered to have both curative and preventative powers, and is at the same time the most potent magic, mystery and talisman. Great secrecy surrounds the Colours, their preparation, and their names.

References to the Seven Colours are said to be found in manuscripts dating from the seventeenth century, and probably in those of the sixteenth century also. They are mentioned principally in the *Awde negist,* existing copies of which are believed by many to be three to four hundred years old. Most texts refer simply to the *säbatu qälämat,* and do not list the colours by name. Where enumerated the lists do not always contain the same seven names, even in the same manuscript. As the use of the *säbatu qälämat* was so well established at this early date it is commonly supposed that the actual substances making up the Seven Colours were all to be found within Ethiopia. The names of some of the colours, however, suggests that they may have been imported, possibly by Arab traders. Although the use of the Seven Colours as a mixture was unknown to mediaeval Arabic medicine, somewhat similar mixtures are to be noted in Arabic folk magic. A common belief is that during the reign of King Zar'a Yakob (1434-1468) the use of the *säbatu qälämat* was abused and that too much evil magic flourished. It is said that the King spoke out against this practice, perhaps trying to prevent the use of the Colours; it is suggested that, as a consequence, they were no longer collected locally but after that date were imported. The later manuscripts tend to emphasise the warning that the Colours are to be used for no evil purpose by beginning recipes: 'In the name of the Father, the Son, and the Holy Ghost. Medicine for the Leg ...' or 'Medicine for the Head ...' as the case might be.

Children are sometimes 'vaccinated' with the Seven Colours as a protection against disease, especially malaria and venereal diseases;

and also against snake bite, crocodile bite, or harm from other wild animals. Mothers may pay an additional fee to have the mixture added to the tattooing colour when crosses on the back of hands, the gums, or 'necklaces' round the throat and chin are to be tattooed. An important use of the Seven Colours is to lend *girma moges*, or dignity, to an individual. Emperors, Kings and other dignitaries are said to have the mixture injected under the skin of either shoulder and learned men to have had it placed in the centre of the forehead. Great men may also wash in the mixture and clean their teeth with *moider* (*Periploca linearifolia*) dipped in the liquid. After the tooth cleaning and washing, the liquid is saved and may be added to *tej* (mead) or *talla* (beer) at banquets. Occasionally the *säbatu qälämat* is used as *mästefaqre*, a kind of love potion, horses even are said to be attracted by persons whose bodies have been washed with the preparation.

The Seven Colours are also applied in educational magic. The use of magic in both traditional church schools and Muslim traditional schools remains widespread, if perhaps rather less so in the latter. This magic serves three main purposes. The first may be termed remedial magic, to treat mental blocks or inattention; the second aims to encourage memory development; and the third is to speed the learning processes, to promote verbal fluency, increase certain manual skills and above all to develop the powers of vision and prophesy. Recipes for this third type of magic contain hallucinogens and sometimes also the Seven Colours. A mixture of this type may be used to write a formula on a piece of sacramental bread, which is administered to aspiring *qene* (poetry) students. For this purpose, the Seven Colours are added to the following three herbs: *Stephania abyssinica*, in Ge'ez called *itse yesus*, 'plant of Jesus': its name is said to be 'written in its roots', and these roots may also be used in recipes for inks; *itse ma'aza* (unidentified), in Ge'ez meaning a 'sweet-smelling plant' having 'three flowers'; and *näç̣ abesho*.

Abesho is actually a term used for any one of two or more species or to a mixture made from these plants. *Datura stramonium* and *Cannabis sativa* are sometimes referred to by this name, though a concoction made from several plants together is held by some Ethiopians to be the real meaning of the term. Such a concoction is used to aid 'intelligence', promote soothsaying or for other magical purposes. Other possible components may be the dried latex of *Lobelia giberroa* and *Biophytum abyssinicum*. The latter, identified from a verbal description, may be included as a magical ingredient because of its 'sensitive' leaves. The term *näç̣* or 'white' *abesho* may allude to one of these plants, or to a mixture of some or all of them; or may mean simply the unripe or 'white' seeds of *Datura stramonium*. Only experienced teachers or *debteras* (deacons) administer *abesho*, known for its adverse side effects. The meal usually taken when these drugs are administered contains *talla* (barley beer), local unleavened bread made from 'white' *teff* (*Eragrostis tef, var. purpurea*), and *nug* (*Guizotia abyssinica*). Milk reduces the effects of the *abesho*, and can if necessary be used as an antidote.

A young man who has since studied in Europe underwent this experience when he was a boy, a pupil in a church school. It was suggested to him by his teachers that he might like to improve both his intelligence and ability as a writer. On the day on which he was to receive his 'treatment', he had to forgo his early morning meal and then sit all day studying in a room filled with books. A number of pieces of iron were in evidence, placed there in order to keep Satan from the books. Holy pictures were brought into the room and hung on the walls: incense was burned so that 'it smelt like a church'. The scholar remembers feeling very hungry as the day wore on. In the evening he was given some special bread to eat, not made of the usual *teff* or wheat, but possibly of *dagusa* (*Eleusine corocana*) the small or finger millet. All night long he shouted and raved, and 'had dreams and saw things'. The boy felt very ill the next morning and it was three days before he had completely recovered. The experience had no noticable effect on his subsequent school work; he never became a writer but did become a painter. Having found it most disagreeable, the boy never repeated the experience and only later did he hear that he had been given the Seven Colours.[5]

The mixture known as the Seven Colours may either be purchased ready-mixed or the different ingredients assembled and prepared by the practitioner. The resulting liquid, if correctly mixed, is supposed to be colourless but the addition of resin may give 'lustre and a golden tone'. The ready-mixed liquid has a strong, sweet smell. It is said that a needle or nail will stand upright in the mixture[6] and that if some of it is poured on to the hand it will roll off, leaving the hand dry, as mercury does. Iron objects are kept with the unmixed colours to ward off evil.[7]

It has already been noted that more than seven substances may be used to make the *säbatu qälämat*. The names enumerated in the different lists do not tally. Here, of course, a certain confusion may have arisen between the name of a substance, the name of the colour or tint produced by that substance, and perhaps its magic name.

In one of the medicinal recipes collected before the Second World War by Gerazmach Gebrewald Aregahegn of Dega Damot the Seven Colours are described as '... green ink, brown ink, cream-coloured ink, shining ink, [red], white and golden ink.'[8] His differs from the list in the modern Amharic-Amharic dictionary, where they are given as 'black, red, yellow, green, blue, flame and gold.'[9]

The following is an early enumeration of the Seven Colours taken from the Ethiopic manuscript Or.11390 in the British Library, which dates from about 1750 A. D. The text, notes and interpretations of the colours are provided by Professor Stephen Strelcyn.

LIST 1

Folio 59ra

(1) *qäyy qäläm* - Amharic 'red colour'. This may also be found with Ge'ez spelling *qäyyəh qäläm*.

(2) *əfran* - 'Saffron', the pollen of *Crocus sativus L.*, probably from Arabic *za'farān*.

(3) *afär mask* (or *məsk*) - 'Musk powder', *afär* is Amharic, *məsk* = Arabic *misk, musk*.

(4) *šäb* - Arabic *šabb* 'alun' [alum] but mostly a mixture of different sulphates, e.g. of Zn and Al.

(5) *ləsanä baḥr* - Arabic *lisan al bahr*, 'tongue of the sea' = cuttle (*Sepia officinalis L.*) - bone, Coptic *las neiom.*

(6) *mästäka* - Arabic *mastakā*, 'mastic, resin from the *Pistacia lentiscus L.*'.

(7) *käfrä lahəy* - (variants *käfarä lahəy, käfära lähay, käfärwälähi*, etc., comes almost certainly from Arabic, maybe from *kaff al-hir* 'paw of a cat', i.e. 'crowsfoot, buttercup', *Ranunculus L.* But this identification remains doubtful.'

Professor Strelcyn goes on to note: 'The same manuscript gives on folio 10v another enumeration of the Seven Colours. Instead of Nos. 4 and 6, there are new terms: *ṭəqur qäläm*, Amharic 'black colour', and *nil* 'a kind of blue', Arabic *nīl* 'indigo', *Indigofera tinctoria L.*' [10]

In an article by Raffaele Cacciapuoti, 'Medicina e farmacologia indigena in Etiopia', *Rassegna di Studi Etiopici*, Anno 1 — Numero III, 1941, he lists six of the Seven Colours.

LIST 2

'... *Lo scioattè calamàt* [Tigrinya for *sob'attä qälamat*] (letteralmente 'i sette inchiostri' in tigrino) rappresenta gli inchiostri (*calamàt*) ottenuti con sette scíoattè sostanze, quasi tutte provenienti dalla lemen, e conosciute sotto i seguenti nomi:

(1) *Gengiàl* - che è un minerale rosso.

(2) *Lesàn bahér* - 'la lingua del mare', polveri ottenute da due differenti conchiglie, una grande e di color madre-perla, e l'altra piccola e rossa.

(3) *Lebsè mariam* - 'vestito di Maria' ottenute da una resina, che stemperata in acqua dà un colore azzurro.

(4) *Afaré mesk* - che si ricava da un vegetale, che quando e secco viene pestato e dà una polvere rossa.

(5) *Zahefràn* - correspondente al nostro zafferano.

(6) *Mesteca seltàn* - che è un'altra resina chi si importà dall'Arabia.[11]

An elderly priest was prevailed upon to show his collection of substances - these being the ingredients of the Seven Colours - which he said had been in his family for thirty years. The names and explanations were provided by him; nothing was taken for analysis and the owner allowed them to be only lightly handled.

LIST 3

(1) *nädd* - Rose-coloured, a shiny, ridged and heavy piece of rock (?) shaped like a broken piece of pencil. The tint was described as *qäyy wärq qäläm*, red-gold colour.
(2) *əfran beča* - Golden or pale yellow irregular pieces, semi-translucent, with the appearance of resin.
(3) *wånde əfran* - Crimson-coloured coral, with the appearance of *Tubipora musica*.
(4) *sete əfran* - Coarse cream-coloured powder or grains.
(5) *näč̣ qäläm* - Pure white powder.
(6) *nil* - Red-brick or brownish pieces, not very heavy, appearing to be a dried substance, or possibly small pieces of brittle rock.
(7) *sete əfran* - Crushed red fragments, which could have been crushed coral or dried vegetable matter such as *Carthamus tinctorius*.

The priest explained that red coral was *wånde*, or male *əfran*; and that white coral was *sete*, or female *əfran*. He added that the crushed substance, in this case red and called by him *sete əfran*, could also be had in white, and that properly speaking here again the male substance should be described as red and the female as white.[12]

Two lists follow, both orally supplied by the same gentleman at different times. The remarks concerning the colours are his.

VERBAL LIST 4

(1) *əfran* - which is red, nearer to crimson than to scarlet.
(2) *kafär* - which is similar to *afärema beča*, earth-coloured yellow.
(3) *mesek* - which is white, looking similar to alum.
(4) *anbär* - which is considered to be black and dark, but which is actually a dun-coloured coarse powder resembling earth, said to 'shine when fresh', and to be found near gold mines. It is said to become black in water. This sample was seen, and had the characteristic strong, sweet scent of the 'modern' Seven Colours.
(5) *kebrä sämay* - which is blue or green. This name is also used to describe crystals of copper sulphate.
(6) *šäb* - which is said to 'resemble water', but actually resembles the colour of a *bora* (palamino-coloured) horse.
(7) *nädd* - which is considered to be 'related to fire', and is the colour of blood-red amber. When mixed with water it becomes red-brown.

These substances are mixed with the juice of the flowers of *dokhma* (*Syzygium guineense*) and with *täkäṣela* (*Loranthus sp.*). which grows on lemon trees. A colourless liquid should result. On being annointed with this liquid, a person will receive all knowledge, dignity, love and respect.

VERBAL LIST 5

(1) *nädd* - It was said to have been brought by pilgrims from the Sudan. It is an old rose-coloured shiny stone, sometimes in the shape of a piece of pencil and sometimes in other shapes. It is dug out of the ground. It is probable that it is not used as a tint but only for its magic properties. It has no smell, though if rubbed on the fingers may leave a very faint odour of milk or butter.
(2) *əfran* - Irregular golden or pale yellow pieces, slightly translucent, said to be the colour of pus. It has a strong smell, something like the smell of blood, but which irritates the nostrils in the way that mustard does.
(3) *hameraye* or *gäbsima* - Crimson-coloured coral. It has no smell, unless very slightly of water.
(4) *zagolema* or *zawl* - The colour is said to be like that of the eye, black and white. It smells like gunpowder, or like a gun recently fired.
(5) *mehaye* - It is pure white, is easily crumbled 'as biscuits are' and has a 'good' smell.
(6) *semeraye* - The colour is 'something that can fit in anywhere', being mixed black, white, yellow and red. It has the consistency of sand or small pebbles. It smells like sulphur.
(7) *alm* - This is black powder. It smells like 'sweaty armpits'.[13]

The following substances were bought in Addis Ababa market, from an aged Arab who dealt in curative, preventative and magic medicine. The colours, available separately, were very carefully weighed out; in addition to the seven substances, the customer was provided with a piece of *muça* or gum and a small bottle of Japanese rose water for mixing them. The price for the Seven Colours ready-mixed was higher. A description of the colours and the names under which they were sold follows below, together with the identifications of the Ethiopian gentleman who provided the two previous verbal lists (and to whom these substances were shown).

LIST 6

(1) *wärq qäläm* - 'Gold colour'. A crimson powder giving a strong stain. Possibly an aniline dye.
Identified as *nädd*.

(2) *wånde əfran* - The coral *Tubipora musica*, lightly dyed, possibly with the foregoing, giving a pale pink.
Identified as *əfran*, though given in his second list as *hameraye* or *gäbsima*.
(3) *məsk* - Small off-white grains, possibly of some resinous substance.
Identified as *mehaye*.
(4) *anbär* - A crumbling brownish substance with a strong smell of spice or incense. Light fawn when mixed with water.
Identified as *hameraye*.
(5) *ṭəqur məsk* - 'Black musk'. Moderately hard black pieces, crumbling fairly easily, with a strong smell of incense or spice or musk. Black in water, but does not dissolve, owing perhaps to insufficient grinding.
Identified as *alm*.
(6) *kafär elai* - Camphor, judging by the characteristic smell.
Identified as *zawl*.
(7) *sete əfran* - Dried *Carthamus tinctorius*, in all probability, with some additives. Giving a cinnamon-coloured stain.
Not identified as one of the Seven Colours.

Asked for by name and purchased in the same shop were a packet of *nil* and a packet of *nädd*. The first appeared to be powdered synthetic indigo, producing the characteristic stain when mixed with water. This substance was identified by the Ethiopian gentleman abovementioned as *semeraye*. The second packet, *nädd*, contained greenish crystals with a tinge of bronze, probably fuchsine, which when mixed with water gave a purplish-red with a metallic sheen. This was also identified as *semeraye*. Both this product and the crimson powder sold under the name of *wärq qäläm* might be described by bookbinders as *bahare arab* - the old term for imported morocco leather. The gum provided with the purchase of the Seven Colours was identified as gum arabic.

Below, the names of the different substances which appear in the six lists are examined one by one, in no particular order.

əfran - Probably from the Arabic *za'farān*, 'saffron', appears in List 1. Named as *zahefràn* in List 2. Named in both verbal Lists 4 and 5, but said to be red in the former, and to be golden or pale yellow pieces in the latter.

əfran beča - 'Yellow saffron', appears in List 3. The name was applied to golden or pale yellow pieces, semi-translucent, with the appearance of resin.

wånde əfran - 'Male saffron', was named in List 3, and appeared to be crimson coral *Tubipora musica*. Was said also to be the name of a reddish vegetable substance. Identified as *Tubipora musica* in List 6.

sete əfran - 'Female saffron', was named in List 3, and appeared to be a coarse cream-coloured powder. Also named in List 3 for a dried vegetable substance, or, less likely, a coral. It was verbally stated by the owner of these two substances, that white coral or a white vegetable substance might be known by this name.

In List 6 a dried vegetable substance, probably *Carthamus tinctorius*, which gives a cinnamon-coloured stain, appeared under this name. When this latter was shown to the author of the verbal lists he denied that it was one of the Seven Colours.

Professor Strelcyn notes that although *əfran* means 'saffron', many different items may be sold under that name. *Wånde əfran* = male plant of *Crocus sativus* and *sete əfran* = female plant of *Crocus sativus*. 'Ethiopian ethnobotany knows many similar distinctions.'

hameraye or *gäbsima* - Appears in List 5, the second verbal list, and is said to be crimson-coloured coral. *Hameraye* may be Tigrinya: *hamar* in Tigre = 'red-brown'. *Hameraye* is also an alternative to *hamerawi* in Amharic, and denotes a light pinky red. It is used particularly as an adjective for silk. *Gäbsima* means 'colour of barley' in Amharic.

afär mask (or *məsk*) - 'Musk powder', *afär* being Amharic and often used for 'earth', and *məsk* = Arabic *misk*, *musk*. This term appears in List 1. In List 2 it is given as *afaré mesk*, said to be a dried vegetable substance producing a red powder. *Mesek* in List 4 was described as a white powder with the appearance of alum. *Məsk* in List 6 was possibly a resinous substance of small off-white grains.

mehaye - Named in the verbal List 5, was said to be a white crumbly substance. The author of this list identified the *məsk* in List 6 as *mehaye* and added that this name is Tigrinya.

ṭəqur məsk - 'Black musk', appears in List 6 and proved to be a black substance smelling of incense, spice or musk.

alm - Was named in the verbal List 5. Said to be a black powder smelling of 'sweaty armpits'. The *ṭəqur məsk*, black musk, in List 6, was identified by the author of List 5 as *alm*.

ṭəqur qäläm - 'Black colour' in Amharic. Named as a variant to the colours in List 1. The colour black occurs in the list in the modern Amharic-Amharic dictionary. The only other use of the word 'black' occurs in List 6 as *ṭəqur məsk*, 'black musk'.

šäb - Arabic *šabb*, 'alun' [alum] but mostly a mixture of different sulphates, e.g. Zn and Al. This name occurs in List 1. Also named in verbal List 4, and said to 'resemble water' but actually to be an off-white palamino colour.

näč̣ qäläm – 'White colour' in Amharic, was named in List 3 and seen to be a pure white powder. White is named as one of the Seven Colours in the earlier enumeration.

ləsanä baḥr - Arabic *lisan al baḥr*, 'tongue of the sea' = cuttle fish (*Sepia officinalis*). This name appears in List 1. In List 2 *lesàn bahér* is described as a powder obtained from two different shells; one large, coloured mother-of-pearl, the other small and red.

anbär - This name occurs in verbal List 4. The colour is considered dark, but a sample seen was a dun-coloured coarse powder resembling earth, which is said to be found near gold mines. In List 6 *anbär* appeared as a crumbling brownish substance smelling of spice or incense. There seems a possibility that this name might be a corruption of *ləsanà baḥr*; or it may be from the Arabic *'anbar* = ambergris.

mästäka - Arabic *mastakā*, 'resin' ('resin from *Pistacia lentiscus*'), is named in List 1. In List 2 *mesteca seltàn* is described as a resin imported from Arabia. Gum arabic was provided by the seller of items in List 6 to be added to the seven other substances but was not included in the Seven Colours.

käfrä lahəy - (Variants *käfärd lahəy*, *käfarä lähay*, *käfärwälahi*, etc.) Almost certainly derived from the Arabic, according to Professor Strelcyn; possibly from *kaff-al-hir* 'paw of cat', i.e. crowsfoot, buttercup (*Ranunculus*). Named in List 1.

kafär - Named in verbal List 4 and described as similar to *afärema beč̣a*, earth-coloured yellow.

kafär elai, *lahay* or *elaye* (the pronunciation was indistinct) - Named in List 6 and proving to be camphor. Arabic *kāfūr* = camphor. The subject of the second doubtful word remains unidentified. This name itself might be a corruption of *käfra lahəy* in List 1.

zagolema or *zawl* - Was named in verbal List 5. According to Professor Strelcyn there are two Amharic adjectives derived from *zag^{wo}al*, 'shell'; they are *zag^{wo}alamma* and *zag^{wo}elemma*, meaning the Italian 'marmorato' and the French 'marbré'. *Zawl* may be an abbreviated form. It was said to be black and white like the eye and to smell of gunpowder. The author of this list identified the camphor in List 6 as *zawl*.

nil - 'A kind of blue'. From the Arabic *nīl*, 'indigo', *Indigofera tinctoria*. Named as a variant to the colours in List 1. Named in List 3, appearing to be red-black or brownish pieces probably of a dried vegetable substance, though they may have been brittle rock. It was impossible to ascertain whether or not this substance could produce a blue.

lebsè mariam - 'Mary's robe'. Named in List 2 and said to be obtained from a resin which, when steeped in water, gives a blue colour.

kebrä sämay - Named in verbal List 4; said to be blue or green. The word is also used for copper sulphate crystals.

gengiàl - Named in List 2; described as 'red mineral'. May be Tigrinya.

nädd - 'Flame' in Ge'ez. Named in List 3; apparently a rose-coloured shiny, heavy piece of rock, ridged like a pencil. In verbal List 4; said to be the colour of blood-red amber and to give a red-brown when mixed with water. In verbal List 5; the description agrees with that used to describe the substance in List 3. Said to be 'dug out of the ground' and to come from the Sudan. *Nädd* is named in the list in the Amharic-Amharic dictionary.

wärq qäläm - 'Gold colour' in Amharic; appears in List 6. The substance so named was a crimson powder, possibly an aniline dye, giving a strong stain. It was identified by the author of the verbal lists as *nädd*.

qäyy qäläm - 'Red colour' in Amharic; appears in List 1 and in both the simple enumerations.

semeraye - Named in verbal List 5; said to be a sandy substance of mixed colouring - black, white, yellow and red - and to smell of sulphur. This may be a Tigrinya word for *semer*, Arabic for brown or brownish. The packet of '*nil*', the dark indigo powder bought in the market, was identified as *semeraye* by the author of verbal Lists 4 and 5.

Faced with a list of between twenty and thirty names, we must suppose either that there are more than seven colours or that each has more than one name. It seems likely that there is truth in both suppositions. A colour may bear either the name of its substance or its tint, the substance itself perhaps changing over the years, as witness the old-rose mineral named *nädd* and the modern dyestuff of the same name. It is said that the colours *nädd* and *əfran* are always included together or separately in 'good' magic; at one time every Ethiopian nobleman was reputed to wear a piece of *nädd* as an amulet to lend him dignity.

As the ingredients of the Seven Colours have always been available to the Church, one is led to speculate on their possible use as painting pigments. One elderly priestly painter described the rose-pink *nädd* as being a beautiful 'old' colour that used to be sold in the market. In preparation for painting it had to be finely ground and mixed with the juice of *qulqwal* (*Euphorbia sp.*) to make it glossy.'[14] *Sete əfran*, used for painting, was generally considered to be *suf* (*Carthamus tinctorius*). *Afär məsk* was said to be a brownish mud-colour and was known as *däma kristos*, Blood of Christ, by 'the painters in the interior'.[15] Another very old church painter was said to keep a small bag of the 'shining' colour round his neck, to allow no one to see it and to use it for painting the faces of the saints.'[16]

There are legends in existence about the use of the Seven Colours in church paintings. In Gojjam some paintings - possibly whole churches - are said to be painted with the Seven Colours. One of these is a picture of the Madonna and the use of the colours is supposed to account for the fact that the eyes in the picture will follow one as one moves, and that people who pass the night in proximity of the painting may see visions or hear oracles.[17] A certain learned church dignitary was appalled at the suggestion that the Seven Colours could be used in religious painting, firmly maintaining that they were employed only in magic.[18] Another dignitary thought that logical reasons existed for all the legends; that, for instance, the swinging censor, producing condensation on an oil painting, gave rise to the idea that Mary wept real tears.

Nonetheless the legends persist. A gentleman, discussing with the author the Seven Colours, regretted that he had never been able to use them for painting in his youth; his father bought the ready-mixed liquid in such small quantities that all was needed as medicine.

If the use in church paintings of the complete prepared mixture known as the Seven Colours would seem unlikely, given that it is supposed to be colourless, that of individual substances comprising it is quite possible.

Mediaeval Arabic medicine employed indigo, cinnabar, red ochre, litharge, sulphur, verdigris and vitriol. It seems not unreasonable to suppose that Arab traders brought into Ethiopia some of these wares; they may have been bought and used as ingredients for the preparation of the Seven Colours. As noted in Part 2, the French physician Poncet observed trade in cinnabar, sublimate and arsenic. In Sennar he also saw '... spica of France [oil of lavender], mahaleb of Egypt (which is a grain with a strong scent)' for sale.[19] Traditionally the Seven Colours are mixed with perfume. It could well be that substances imported for medico-magical purposes provided a source of materials for painters. Certainly the early foreign artists would immediately have recognised, and wished to make use of, pigments already familiar to them.

As a result of an analysis carried out on some Ethiopian paintings just before they were loaned for an exhibition of religious art in 1973, Fritz Weihs noted that the principal pigments used were:

(1) Cinnabar (red mercury sulphide), 'to be found in the north of the country'.
(2) Orpiment (arsenious trisulphide), 'to be found in the east of the country (Dalol)'.
(3) Green - a mixture of orpiment and indigo.
(4) Indigo - 'probably through trading with India, later produced in the country.'
(5) Charcoal.
(6) White gypseous chalk - 'to be found in the country'.

Green earths were found in some paintings; also a red lacquer ('presumably madder').[20]

An advisor to the Ministry of Mines, some ten years in the same post, expressed himself as 'very surprised' to hear that cinnabar was to be found in Ethiopia, stating that the Ministry of Mines possessed no such information.[21] A senior geologist at the University of Addis Ababa, however, thought it was possible that small pockets might exist. The same geologist agreed that orpiment was to be found in the east, in Dalol, though we should not forget that Poncet, in the West, saw it imported from Sudan.[22]

With regard to indigo being 'later produced in the country', information is somewhat confused. There are no references to the cultivation of indigo by the Portuguese Jesuit Mission. As they maintained contact with their Order in Goa it was easy for them to import it from India.[23] Viscount Valentia in 1809 wrote of carpets being dyed with a plant resembling indigo.[24] Johnston, writing about 1840, tells us that King Sahle Selassie expressed the wish that indigo should be cultivated and his people taught how to prepare the dye;[25] but von Heuglin, describing his journeys in 1861-62, saw indigo growing wild and observed that it was never used.[26] Bearing in mind that *nil* appears as one of the Seven Colours, perhaps it was always customary to import it.

In some paintings in which the pigments were analysed green earths were found. There are several known areas where these are easily available; doubtless there exist a number of pockets. One likely source is Azezo, 11 km. from Gondar; and there are also coloured clays considered to be worth exploiting commercially between Debra Berhan and Debra Sina.

Only the pigments used in certain of the paintings were analysed. We do not yet know what substance was used for a pale azure or cerulean blue to be found in the *haräg* in the early manuscripts and also in some church mural painting. It has long been considered probable that the ancient Egyptians mined copper where malachite and azurite are to be found in Eritrea: the Portuguese later noted these ancient workings, digging one pit and some galleries at a place called 'north hill' at Adi Rassi.[27] Azurite was thus available to the early painters; if it be true that originally only indigenous substances were used it may have provided blue for the Seven Colours. Similarly malachite from the same mines could have provided a source for the colour green. Alternatively the pale blue in the *haräg* might be one of the blue earths of which there are small pockets, or even verdigris.

If we take a look at the other colours used by mediaeval painters in Europe we may see how many were available in Ethiopia.

Sinope is the ancient name for all native red iron oxides. Bole, the old name for which was poliment, is widespread in Ethiopia, existing as clays coloured by iron and also in a harder form, which is similar to Pozzuoli red. The monks at Lalibela report that a very local supply, made use of in times past, is now exhausted. A very brilliant red found near the monastery of Debra Libanos is known to have been used seventy years ago. Cinnabar has been mentioned; while folium, the ancient mulberry colour of various vegetable origins, now generally known as lakes, was employed by some late fifteenth-century painters. Ethiopian red inks which were used for calligraphy on the parchments are, when soil is added, a form of lake; and although *Rubia tinctorum*, from which was made madder lake, is not native, *Rubia discolor*, African madder, is used both as a dye and in inks. Fritz Weihs suggests that the madder analysed in some paintings might have been imported by the Portuguese. Lac, of course, was imported in very early times and the practice may have continued for centuries. Lac, cochineal or a similar insect dye, kermes, was probably brought in to Harrar. Keller mentions imports of cochineal in the nineteenth century.[28] We know that it was the custom in Harrar to import red dyes. In Europe a fugitive red lake used to be made from *Carthamus tinctorius* under the name of Carthame. This was readily available. Orpiment has been mentioned; with it must be linked realgar, the arsenic orange. Ochre, native clay coloured by iron oxide, is widespread. *Terra merita*, a fugitive yellow lake made from saffron or curcuma, could have been used. *Curcuma tinctoria* does not grow in Ethiopia; but *Curcuma longa* is available and saffron could have been imported as one of the Seven Colours. Other plant lakes were derived from berries and barks. Barks were used as dyestuffs by the monks and berry extracts included in the inks.

It is with the pigments once known as sinope and poliment, with the different ochres, chalks, and the blue and green earths, that the walls and ceilings of rock churches are decorated. In some of the paintings black outlines and the black pupils of the subjects' eyes have disappeared, leaving a curious impression like that of a photographic negative. This may be seen in Tigre in Arba Itu Insessa church and also in Qorqor, in Mariam church on Qorqor mountain. The reason the black colouring has fallen out is almost certainly because it was made with a mixture of charcoal and roasted grains, as used in the inks. The *mashega*, animal skin glue, or *muça*, vegetable gum, used in the preparation either did not retain its holding properties or caused excessive binding: the black aided by time and damp, has flaked off.

Not infrequently the colours appear to have been applied upon very inadequate priming; the walls were, in effect, simply stained by the coloured earths. The greater part of the wall paintings are applied over a type of plaster, which in general has held up well when damp is minimal. Nowhere does there appear to have been any elaborate waxing or protective coatings over wall paintings. In some of the paintings in the monolithic church of Gannata Mariam most of the ochres have vanished, leaving only black and a pale cerulean blue. In the places where the earth colours failed to hold (probably owing to damp) the compound substance making up the black was presumably well mixed with glue or gum. We may wonder whether the pale blue was a powdered azurite rather than an earth, and was similarly treated. Further analysis of the ancient pigments remains to be carried out: there are still many churches whose treasures have not yet been fully examined or photographed and where, alas, they are suffering decay and even destruction.

Many languages are spoken in Ethiopia; even within one language group it is not easy to identify dye plants from the name alone. Very often this is due to the fact that the plants are also used in traditional medicine. As Harris notes: 'The physician's lore is kept a profound mystery ...'[29] Not only may a single plant be given several names, possibly to mask its identity, but a single name may be used for several plants. It is possible for a herbalist, visiting a district somewhat removed from his own and asking ten herbal practitioners of that district to bring him a certain plant, to be presented with no less than ten different varieties, which are known to those men by the same name. He will later be assured that these plants give identical results in medicine.[30]

A further difficulty in identifying plants in Amharic stems from the fact that in the past there have been virtually three languages in use. Firstly the language of the Church, where many terms remained in the original Ge'ez, secondly the language of the educated aristocracy, who probably spoke the purest Amharic; and thirdly the language of the ordinary people. Such differences are reflected in plant names; the church designations for important medicinal or colour-giving plants may be in the ancient Ge'ez and the common name corrupted to Amharic. An example of this is *Datura stramonium*, which is known by the clergy as *itse fars*, 'plant of Persia'. In Amharic the plant is both *attefaris*, a corruption of the above, and *astenagher*, possibly a different species but possibly also a reference to the ripeness of the seed, differentiated by the name because of consequent different properties, the word meaning 'which makes one talk too much'.

The book known as *Arde'et*, the magical Book of the Disciples, tells how Jesus taught the disciples His secret names. The belief in the power of the name is strong, even today; many people are known by names which are not their real ones, real names and Christian names being 'secret'. Men may change the names by which they are commonly known, rather in the manner of one wishing to change his luck; this in order to confuse evil spirits and to counteract the charms of magicians.

Knowledge used to be considered sacrosanct, of far too great a value - and possibly too dangerous - to be disseminated among the unworthy. In effect, part of the value of knowledge lay in the very fact that it was a private and esoteric thing. As in some other countries, there existed a belief that if a name, for instance of a drug or herb, were known or spoken, the virtue would go out of it; it would cease to cure. To know the true name of something was to have power over it. So men went alone to cull their herbs, even members of their own family having to turn aside when a herb was identified and plucked. Thus was secrecy preserved.

The following recipes present what must surely be only a small part of the rapidly vanishing colour lore extant in Ethiopia today.

REFERENCES AND NOTES PART 3

1 *Aleqa* Imbakom, verbal information.
2 Chojnacki, S., 'Some notes on the history of the Ethiopian flag', *Journal of Ethiopian Studies* Vol. I, No. 2.
3 Mercier, Jacques, verbal information.
4 Gilbert, Michael, verbal information. [This reference number is missing in the chapter text of the 1986 edition. Michael Gilbert was the Director of the University Herbarium in Addis Ababa.]
5 Kessela Markos, verbal information.
6 Megabi Tsaga Flate, verbal information.
7 Yegezu Bisrat, verbal information.
8 *Journal of Ethiopian Studies* Vol. IX, No. 1, p. 145.
9 Dästa Täklä-Wåld, *Addis yamarañña mäzgäbä qalat*, 1970, p. 1160.
10 Lists and notes provided by Professor Stephen Strelcyn in correspondence.
11 Information supplied by Professor Strelcyn in correspondence.
12 Marigéta Tsege, verbal information.
13 *Aleqa* Imbakom, verbal information.
14 *Aleqa* Desta, verbal information.
15 Alemayehu Moges stated verbally that *afära məsk* and *aferma* mean the same thing. Baeteman (col. 641) gives *aferma* = *couleur de cendre*.
16 *Aleqa* Yohannes.
17 Alemayehu Moges, verbal information.
18 *Aleqa* Desta, verbal information.
19 Poncet, C. J., op. cit., p. 107.
20 *Religious Art in Ethiopia*, 1973, p. 303.
21 Dr Hamrla, verbal information.
22 Dr Morton, verbal information.
23 Dr Mered Wolde Aragai, verbal information.
24 Valentia, G., op. cit., p. 162.
25 Johnston, C., op. cit., p. 418;
26 Heuglin, M. Th. von, op. cit., p. 328.
27 Jelenc, D. A., *Mineral Occurences in Ethiopia*, 1966, p. 224.
28 Keller, op. cit., p. 36.
29 Harris, W. C., op. cit., II, Appendix.
30 Alemayehu Moges, verbal information.

The Recipes

GREEN

Adhatoda schimperana	SEMIZA, SENSEL (A)
Calpurnia aurea	DEGETTA, DIGITTA (A)
Canavalia gladiata	ADENGUARI (A), HEPO (G)
Capparis sp. prob. *tomentosa*	ANDEL (T), GUMERO (A)
Catha edulis	CHAT (A)
Datura stramonium	ATTEFARIS, ASTANAGHIR (A)
Dombeya goetzenii	SHAWUKO, SHUKO (K)
Euclea schimperi	DEDEHO (A), KILLIAW (T)
Macaranga kilimandscharica	SHAKERO (K)
Maesa lanceolata	KELEWA (A)
Phytolacca dodecandra	ENDOD (A, T)
Rhamnus prinoides	GESHO (A)
Rhamnus staddo	T'ADDO (A), TZEDO (T)
Zehneria scabra	AREG RESA (A)
Unidentified moss or lichen	YE DINGAI SHEBET (A)

Adhatoda schimperana - SAMIZZA, *SEMIZA*, SMISA, SANSAL (A), TIMISA (A-Gojjam), DUMOGA (A-K), SHIMFA (T)

S. Edwards gives Sensel (A, Shoa), Sim'iza (T).

Recipe: Take SENSEL or SEMIZA leaves and squash them with water. The colour is said to be fast.

Sources: Yacob Kassahun, who gives the colour as fast.
Wondimagegn Tefera (Illubabor).

Calpurnia aurea - DEGISSA, DIGITTA, ZIGITTA (A), MISSIRIK (A-Gondar), HEZAUS, SITARA (T), CHEKA, CHEKETTA (G), SOTILLA (G-W).
East African Laburnum.

Lemordant gives *zəgəṭṭa*, *dəgəṭṭa*, and *dəgiṭṭa*.

Breitenback gives DIGHITTA, MISSIRIK (A), SOTELLU (G), SITARAH, HESANTZ (T).

Note: Breitenbach gives DIGHITTA (A) for *Sesbania punctuata.* Mooney gives DIGITTA as (A-Gondar) for *Carissa edulis. Ato* Lakew of the Science Faculty of H.S.I.U., thinks that this is incorrect and should probably refer to *Syzygium guineense*, which Mooney lists, among other names, as DIGHITTAN (A). However Wondimagegn Tefera, one source below, gives both DOKMA (A) and BEDESSA (G) (*Syzygium guineense*), for blue using the fruit, and DIGITA (A) as a different plant for obtaining green.

Recipe: Pounded DEGETTA leaves give green.

Sources: *W/t* Azeb Desta (Harrar) gives the recipe above.
Wo Tsehai Haile (Addis Ababa), suggests adding a little water.
Lemma Arerru (Gemu Gofa) uses the spelling DIGITTA.
Wondimagegn Tefera (Illubabor) gives DIGITA (A), which he describes as a 'herb'. No recipe.

Used in the recipe for Black Ink of *Abba* Wubu Kidane Mariam.

Canavalia gladiata - ADENGUARI (A), *HEPO* (G).

HEPO (G) is described as a domestic plant, a slightly bitter type of bean of medium size, speckled grey/black/brown.

Recipe: Take HEPO leaves, cut them into small pieces and add a little water. Squeeze the wet leaves. This is used as 'green ink' by 'churchmen' and for dyeing the edge of the cloth during weaving.

Source: Shiferaw Fufa (Wollega).

Capparis sp. probably ***tomentosa*** - GIMERO, GUMARU, GUMURO (A), ANDAL, ANDEL, ANDELO (A, T), GOMBOR (Harrar).

see also Red, Black and BLACK INK.

Recipe: The fresh leaves of GUMERO can be crushed and used for green.

Sources: *Aleqa* Gebre Selassie (Selale) and
Aleqa Wolde Medhin (Menz).

Catha edulis - CHAT (A), GOFA (G), KAT (G-H), QAT (Ar), *CHID, GELEBA* (Ge'ez).

Recipe: The fresh green leaves, pounded, are sometimes mixed with a little water. The resulting light green liquid is moderately fast.

Sources: Debebe Haile Giorgis (Harrar), who gives the recipe above. *W/t* Azeb Desta (Harrar).

Datura stramonium - ASTANAGHIR, ASTENAGRT, ATAFARIS, *ESSEFARIS* (A), *ITSE FARS* (Ge'ez), *ASANGIRA* (?), *MENGI* (G), *MESTANAGER*, *MEZERBOL* (T).

Cufodontis gives: ATAFARIS, ATTAFARIS, EZAFARIS, ASTANA GHIR, STANAGERT (A), ASANGHER, ASAN-GRA (Goromo), BOA-MADOW (Som.), MASABAH, MESERBA, MESERBAH (T), THIMFRAH, THIRUFRAH, THRIFRA (Tigre Mensa). He adds that oil from the seeds is used as a liniment for pain and steam from the leaves as a narcotic.

The plant is known by the clergy by the Ge'ez name ITSE FARS, literally 'Plant of Persia'. The clergy assert that there are two species of *Datura*, known in Amharic as ATTEFARIS and ASTENAGHER, the latter meaning 'which makes one talk too much'. Lemordant also surmises that there are probably two species of the same type, this accounting for the different names. He quotes Griaule, who, writing of ASTENAGER, says: 'il ressemble au datura stramonium. Il existe deux espèces, l'une à fruits noirs, inutilisée, l'autre à fruits blancs que les étudiants emploient.'

A species somewhat similar to *D. stramonium* is *D. metel* which usually grows at a lower altitude and is slightly less toxic. However it seems possible that the different Amharic names may be attributed to ripe or unripe seeds, the unripe being referred to as 'white' and the ripe as 'black'. *Datura stramonium* seeds are used in magic under the name *Abesho*, or sometimes *näč̣ abesho*, white abesho: the latter may refer to unripe seeds.

Recipes

(a) ATTEFARIS, ASTENAGIR. Pound the fresh leaves. According to some sources the leaves should be mixed with water. Several sources mention that the colour becomes 'stronger' if mixed with soot. There is a difference of opinion as to whether or not the colour is fast.

(b) Take the ESSEFARIS fruit, grind it, and then stir with warm water. (If fruit is first roasted, the colour will be black. See under Black.)

(c) See recipe (c) under *Euclea schimperi* for DODOHO and ESSEFARIS mixed together.

Sources: Numerous for both recipe (a) and recipe (b), coming from Arussi, Bale, Begemeder, Gojjam, Sidamo, Harrar, Eritrea, Tigre, and Illubabor.

Yacob Kassahun gives MEZERBOL (T).

Tsegahun Mebrahtu (Tigre) gives MESTANAGER (T).
Haile Michael Kassaye gives MENGI (G), saying that this is the Gallinya for ASTENAGHER.
Wondimagegn Tefera (Illubabor) gives ASANGIRA (G).
Cufodontis gives somewhat similar names in Goromo.
(Mooney gives *Millettia ferruginea* as ASANGERA (G-W).)

Dombeya goetzenii - WOLKAFFA, WULKAFFA (A), ULKIFFA (G), DANESSA (G-K and G-Bale), SHAUKO, *SHUKO* (K).

A tree. Given also by source below with the spelling SHAWUKO (K) under Brown.

Note: Sister Albertus of Our Lady of Mercy T.T. Centre, Mbooni, Kenya, in *Art Teaching for Primary Schools in Africa* by J. C. McKenzie, OUP Nairobi, 1966, lists the Kamba name MUKEU for making green. This MUKEU was identified by Dr J. O. Kokwaro of the Botany Department of Nairobi University as *Dombeya goetzenii*. Sister Albertus gives the following recipe for green: 'The stems of this plant are boiled'.

Recipe: Take the uncooked leaves of the tree SHUKO (K) and squeeze them to 'decorate furniture'.

Source: Bekele Wolde Gabriel (Kaffa, Bonga).

Euclea schimperi - DEDEHO, *DODOHO*, MIECHA, MIESA (A), KELLAU, KILLIAW (T, ?A), GUM (T-Mensa), DUBOBIS (Harrar), DADAHO (G-H), KANKO (G-Sid), HANKU (Gugi), DUPPO, GINO (K).

Breitenbach gives: DEDAHO, KURKURA (A), MIHESSA, GHINO (G), KELLAU, GUM (T), DUBOBIS, MAYER (Som). The fruit is edible: under the name ZIETO or ZEYATO (T), it is used as a blue dye.

Lemordant quotes Guidi as describing *dädäho* as 'espèce de petit arbre ressemblant au genévrier', Juniper tree, adding that shepherds and children eat the fruit and that an infusion of the roots is used medicinally.

Note: For *Myrica salicifolia* Mooney gives: BAREDDU, DEDECHO, KALAWA, SHINAT, SHINET (A), NEVI (T), ABAAI, KATABA, RADJI (G); Breitenbach for *Myrica salicifolia* = *kilimandscharica* gives: SHINET, SHUNATZ, DEDAHO, BARODDU, KALAVA (A), ABAG, KATABA, RADJI (G), NEBI (T), adding: 'The young leaves, dried, pulverized and mixed with butter, are used as a house medicine against skin diseases.' There is a possibility that DEDEHO mentioned under *Euclea schimperi*, and KALAWA, under *Maesa lanceolata*, both refer to *Myrica salicifolia* and that this is one of the many examples of the same name being used for more than one species, particularly in cases where plants have medicinal significance. *Salvadora persica* is another plant with a similar name for which Mooney gives: DADAHO, HADAI (A), ADDAI (T).

Recipes

(a) When there are fresh green SEEDS of KILLIAW (T), they may be picked, crushed, and placed in a container to which a little water has been added. This gives 'green ink'.

(b) Take the ROOTS of DEDEHO (A), chop them and boil with water. When the water is nearly cold, dip in the cloth. Fast colour.

(c) Take DODOHO (A) fruit with ESSEFARIS (A) (*Datura stramonium*) fruit. Grind both and stir with warm water.

Sources: Assefa Asghedom (Kaffa), gives recipe (a) from experience in Tigre. He adds that not all plants have seeds (?).

Taddessa Wolde Meskel gives recipe (b) for green.

Ato Imeshaw Temtin (Bulga, Shoa), gives recipe (b) but the resulting colour as brown.

Ato Kessela Markos (originally of Gondar), describes the roots of DEDEHO as becoming red-brown in water.

Mekuria Degu (Shoa), gives recipe (c) for green.

Tsegahun Mebrahtu (Tigre) gives YEDEDEHA ABEBA (flowers), but mentions no colour. It is probable that seed heads were intended; the flowers are white.

Gebre Yohannes (Tigre), mentions KILLAU for green, but gives no recipe.

Macaranga kilimandscharica - SHAKARO, *SHAKERO* (K).

Recipe: Take the leaves of SHAKERO and boil with water. Gives a green colour for 'decorating furniture'.

Source: Bekele Wolde Gabriel (Kaffa) describes this and other dyes as used by the people of Bonga, Kaffa.

Maesa lanceolata - ABALIYEH, AKALUA, ALGALUA, KALAUA, KELEWA (A), GESHI (G), ABAEI (G-W), ABAIYEH (G-Sid), IMBIS (Wollamo), CHAGO (K).

Breitenbach gives: KALAWA, AKALUA (A), ABBAYE, GESHY, IMBIS, GERGESHO (G), SOAREA (T), ARAR, MAAS (Ar) and adds that 'the fruits yield an oil which is used as a grease for pottery and as a vermifuge'. The slash exudes drops of dark red to brown resin. This resin exudes from the veins when the leaves are broken across.

Recipe: 'The seed and leaves of KELEWA serve the same purpose and are prepared by the same method as ETSEFARIS.' (As this reference is to *Datura stramonium* it is presumed that the ensuing colour is green, despite the red-brown slash of bark and leaves.)

Source: Alayu Gorfe (Harrar).

Phytolacca dodecandra - INDOT (A), ENDOD, ENDUET (A, T), MAKAN-ENDOT, SHEBTI, SHIPTI, SOBETH (T), ENDODA (G).

'The bark and roots are very poisonous, the leaves and green fruits rather less so, while the ripe fruit appears to be harmless in small quantities.' Contains saponin, a fat, and tannic acid. (*Common Poisonous Plants of East Africa*, Bernard Vercourt and E. C. Trump).

Cufodontis mentions the use of the roots for venereal diseases, the leaves for worms and wounds, the seeds for worms and the cooked fruit for soap.

Griaule gives ENDOD as *Phytolacca abyssinica* and says: 'on l'utilise également dans la fabrication de la couleur jaune.' To date no recipes have been received which mention this use of the plant for making yellow.

Recipes
(a) Press the ENDOD leaves without adding any water.
(b) The pounded yellow seeds of INDOD give white colour.

Sources: Arega Gebre (Kaffa) mentions that ENDOD may be mixed with soot to produce a black ink. The leaves are also used with powdered charcoal to repair blackboards in schools.
Hapteselassie Geleta (Wollega), among many others, mentions the widespread use of ENDOD as a soap.
Yacob Kassahun asserts that the green colour is fast.
Endale Taddesse (Illubabor).
Temam Ibrahim (Illubabor).
Worku Ambie (Gojjam).
W/t Azeb Desta (Harrar) gives recipe (b). It is not certain whether this is intended to mean bleach. Most berries are scarlet.

Mentioned in the recipe for Black Ink No. 3, of *Memhere* Hailu Man Wasinot.

Rhamnus prinoides - GESHO (A), EHK (A, T).

Breitenbach gives *Rhamnus prinoides* ≡ panciflorus as GESHO, GHEBO, EK (A, T). He also notes the use of the name GESHO (A) for *Zizyphus mucronata*.

Historical note: M. Th. von Heuglin in *Reise nach Abessinien*, 1874, mentions '... Tado und Geso [*Rhamnus panciflorus* and *R. staddo*], deren Wurzeln und Blätter zur Konigwein Bereitung benutzt werden.' Gesho is widely used for making *talla*, a beer, and is by no means confined to flavouring the 'royal wine'. The seeds, YE GESHOFERE, are given as producing yellow.

Recipes

(a) GESHO leaves are crushed to produce green.

(b) GESHO is boiled with salt for a light khaki green. The cloth should be dried in the sun but turned every ten minutes or so; the colour becomes too dark if exposed to too much light during the drying process. A yellower colour is obtained when the cloth is dried in the shade.

Sources

(a) From 1968 notes, source name lost.

Ato Imeshaw Temtin (Bulga, Shoa), gives recipe (b) and describes GESHO as 'one of the best colours'.

Rhamnus staddo - *T'ADDO, T'EDDO* (A), *TZEDO, TZODO* (T).

Mooney gives *R. staddo* as THADDO (T).

Lemordant gives *R. deflersii* for the same plant.

Historical note: In *Reise nach Abessinien*, M. Th. von Heuglin, 1874, 'TADO' is mentioned as *Rhamnus panciflorus*, '...deren Wurzeln und Blätter zur Konigwein Bereitung benutzt werden.' It is used extensively to flavour *tej*, honey wine, not for the 'royal wine' only.

Note: Ogbai Zeru (Eritrea), reports the use of TZEDO (T). 'It is used in *mes* [the Eritrean *tej*], to make it yellowish and strong. People strip the bark for miles around Asmara as it is very expensive to buy in town.' The fruit is used to dye violet.

Recipes

(a) The bark of TZEDO is dried and then soaked in water for about five days, the water turned yellow for dyeing.

(b) To one litre of water add a handful of T'ADDO twigs, complete with leaves. Boil at 100-120° C and add 30 grams (1 tablespoon) of alum and the same of salt. Put in the wool and continue to boil until a good deal of water has evaporated. Take off the fire and leave for 24 hours before removing the wool. Gives a soft olive green.

Sources: Hadgu Bariagaber and Ogbai Zeru (Eritrea) give recipe (a). Worede Isaac (Eritrea) gives the pronunciation TZODO (T) and says the 'stems' should be used to give yellow.

Sig. Giraldi and *Ato* Tekle Haile of the Empress Menen Handicraft School Dye Shop give recipe (b). Sig. Giraldi has used it only to dye wool, but says that he made extensive use of it in the days before the importation of chemical dyes.

Zehneria scabra - AREG RESA, *HAREG RESA* (A), *ITSE SABEK*, from the Ge'ez *eṣ̂ä sabéq*, used extensively by the clergy. HAFA FELO, HAFFAFALA, HAFFA-FALU, HASFAFALA (T), *HIDDA FITA* (G-Shoa), *UMBAO, IMBAU* (G. Wollega and G. Illubabor).

The above names are taken from Cufodontis with the exception of those in italics.

Lemordant adds that students drink the juice to obtain 'science et éloquence'.

Recipes
(a) The leaves are pressed and no water is added.
(b) The leaves and the thin bark are used to make green. Add the seeds also 'if you can find them'. Dry and powder these three constituents and add a gum medium when the colour is required for painting.

Sources: Worku Ambie (Gojjam) gives recipe (a).
An old man in the School of Fine Art, Addis Ababa, gives recipe (a).
Aleqa Gebre Medhin (Menz) gives recipe (b), using the name ITSE SABEK.
Negaso Gidade (Wollega) gives UMBAU and IMBAU as being the G. Wollega and G. Illubabor for this plant.
Adane Feyyisa (Bale) gives HIDDA FITA as being G. Shoa.

A moss or lichen - YE DINGAI SHEBET.

This moss or lichen, pale in colour, is found on stones and is named after grizzled hair.

Recipe: Collect YE DINGAI SHEBET and then dry and pound it. It can be used to make green.

Source: *Aleqa* Wolde Medhin (Menz).

YELLOW

Acacia abyssinica or *A. seyal*	WACHU (A)
Bidens spp.	ADDAI ABEBE, YEMESKEL ABEBE (A)
Croton macrostachys	BISANA (A)
Curcuma domestica	ERD (A), ERDI (T)
Echinops sp.	DANDER (T)
Hagenia abyssinica	KOSSO (A)
Helianthus annuus	SUF (A)
Prunus africanus	KEMA, EKEMA (A)
Rhamnus prinoides or *Zizyphus mucronata*	GESHO (A)
Rumex abyssinicus	MEKMOKO (A)
Rumex steudelii	TULT (A)
Solanum spp.	IMBWAI (A)
Terminalia brownii	WEIBA (A)
Terminalia glaucescens	ABALO (A)
Unidentified, ?*Nuxia congesta*	CHOJO (?A)
Historically noted: Berberis sp.	

Acacia abyssinica - GRAR, BAZRA-GRAR, WACHU (A), ALLA, CHIAA, NEFASIA (T).

Breitenbach gives: GRAHR (A), LAFTU, GARBI (G), DADECHA (Arussi). A high altitude acacia with brown fissured bark and pale red slash.

OR

Acacia seyal - SEYAL (Ar), WACHU (A), WAIKU (G), WAJO (G-H), CHIUA-KAIYEH (G-Sid), WAKO-DINO (Boran), MAKANI (Danakil). A low altitude acacia. The bark contains 18—20 per cent tannin and yields a red liquid extract.

Note: Acacia caffra is known as African cutch. Cutch was well known in Europe as a dye stuff.

Recipes

(a) The bark of WACHU is ground or chopped and boiled with water to which salt has been added. Fast colour.

(b) The twigs and leaves are boiled and the colour can be used to dye hides.

Sources: Mekuria Degu (Shoa), gives recipe (a).

Ato Tekle Haile, Empress Menen Handicraft School Dye Shop, gives recipe (b).

An Ethiopian gentleman visiting the School of Fine Art, Addis Ababa, mentioned that many people dyed their clothes with WACHU during the Italian war and occupation. It gave an excellent camouflage 'khaki' colour.

Bidens borianiana, Coreopsis borianiana: the Maskel daisy. Also ***B. chaltodonta, B. macrantha*** and ***B. prestinarieaeforrnis*** – ABEBA-MESKAL, ADEY ABEBE (A).

S. Edwards gives Adey abeba (A), Kéllo (G. Illubabor, Shoa), Gelgele meskel (T).

Lemordant gives *Coreopsis macrantha* as ADAY ABABA and adds: 'on utilise aussi le pigment jaune obtenu par decoction pour les peintures artisanales artistiques locales'.

The flowers are also used in the recipes for Red Ink.

Recipes

(a) Simply rub the petals of YEMESKEL ABEBE on paper.

(b) Collect the ADDAI ABEBA. Dry and winnow them, and then pound and grind them to powder. This powder is then mixed with water and used to dye cloth, for decorating, and for painting pictures. Reputed to be a fast colour.

(c) Collect the flowers and put them into a container with a little water.

(d) Pure white ash from wood is added to YEMESKEL ABEBA when dyeing grass for basket work.

Sources: Simme Bekele (Arussi) gives recipe (a).

Bayelegne Meshesha (Bale) gives recipe (a).

Taddesse Woldemeskel gives recipe (b).

W/t Azeb Desta (Harrar) gives recipe (b).

Mulat Admassu gives recipe (c).

W/t Yenagu Dessie (Gojjam) gives recipe (d) and adds that children mix ash and flowers on their hands to dye them red. She is uncertain whether the recipe dyes the grass yellow or red.

Numerous other sources from Sidamo, Tigre, Illubabor, Gojjam and Shoa.

Croton macrostachys - BASANNA, BISANA, MESANNA (A), AMBUKH (T), TAMBUK (T-Tigre), BAKANICHA, ?MOKONISSA (G), BAKANISSA (G-W), MASSAGANTA (Konso), WAGO (K), MASINCHU, WUSH (Sb).

Cufodontis states that the fruit and roots are decocted as a medicine for venereal disease and the bark, crushed and mixed with KOSSO (*Hagenia abyssinica*) for worms.

Recipe: The dried leaves of BISANA are pounded and the powder used for a yellow dye.

Source: Wit Azeb Desta (Harrar).

Curcuma domestica, Turmeric — URT, ERD, *ERID*, *ERED* (A), *ERDI* (T), HARD (G-H), *HOORID*, *HURUD* (Ar or T).

According to Purseglove, the name for turmeric in Sanskrit and some other languages is synonymous with yellow. With the addition of lime or other alkalis the rhizomes may also produce a red or red-brown dye. Although the dye may be used without the addition of a mordant, the colour is fugitive.

Note: It is possible that some sources intended ERET (A), *Aloe spp.*, whose flowers are also occasionally used for yellow.

Recipes

(a) Powder the root and dissolve in boiling water. Then put in the cloth to be dyed.
(b) As above, but adding equal parts of oil and water.
(c) For use on paper simply dampen the powder, as with powder paint.

Sources: Altaye Gizaw gives recipe (a) and says the colour is fast.
Arefa Gebresadik (Tigre) gives recipe (a).
Abdulkadir Ahmed (Eritrea) gives recipe (b), says the colour is fast, and gives the Arabic as HERID.
Simme Bekele (Arussi) gives recipe (c) and the name ERID.
Idris Osman Samra (Eritrea) says the colour is fast if the cloth is well boiled, giving the names HURUD and HOORID.
Ashagrie Tasew (Bale) also says the colour is fast, and gives the name ERID.
Yehdego Melles (Tigre) gives ERDI (T).
Yacob Kassahun also gives ERDI (T).
Worku Ambie (Gojjam) simply says 'use dry'.
Eshete Degefu says 'simply rubbing it'.

Other sources include three from Illubabor, three from Tigre, three from Eritrea, two from Sidamo, and one each from Begemeder, Gemu Gofa and Hararghie.

Echinops sp. - DENDUR, DENDURRA, KOSHISHILA (A), DANDER (A, T), KABARITCHO (A).

S. Edwards gives Kwasheshila (A), Dandér (T).

Recipe: Fresh DANDER rubbed on paper gives a clear yellow.

Source: Bissrat Fesahaye (Tigre).

Hagenia abyssinica - KOSSO (A), HABBI (T), DUCHIA, EDO, FIETO (G), HETO (G-K).

Breitenbach gives: KOSSO (A), DUSHA, FIETO, HETU (G), HABBI (T).

KOSSO is a well-known vermifuge, especially for tapeworm, and is prepared from the dried female inflorescences. The wood is dark red to red-brown, medium soft.

Recipe: Cut and pound KOSSO wood and mix with water and mud in a hole in the ground. Bury the article to be dyed in this mixture for four days.

Source: *Ato* Tekle Haile, The Empress Menen Handicraft School Dye Shop.

Helianthus annuus, the sunflower - SUF (A).

Recipe: Dry and pound the petals of SUF. Add a little gum resin and use with yolk of egg for painting.

Source: *Ato* Kessela Markos, whose father was a noted calligrapher in Gondar, used to use this recipe for yellow when a boy.

Prunus africanus - AKOMA, AKWOMA, *KEMA*, TUKUR-INCHET (A), BOSOKO (A-Wollamo), MOKA-DIMA (G), GARBEH, GARBI (G-W), HOMI, OMI (G-W, G-K), ME-ESA (G-K)', BERU (Chako), SOSCHE (Sh).

Cufodontis adds MICHICHIO (Sid).

Lemordant gives akama, əkma, and ahoma (T), describing the tree as looking similar to a European Oak.

Ato Alemayehu Moges describes the trees as growing near Gondar, and gives the following pronunciations: *KEMA*, *KIMA*, *EKEMA*.

Recipe: Take the bark of KEMA and boil. Gives a khaki yellow to cotton.

Source: *Ato* Tekle Haile, The Empress Menen Handicraft School Dye Shop.

Rhamnus prinoides or ***Zizyphus mucronata*** - GESHO (A)

Rhamnus prinoides has a red shiny berry, up to ¼ in. long. See Green.

Breitenbach gives *Zizyphus mucronata* as GESHO (A), for which Mooney gives GUECHO (A), GEBA, GHEBBA (A, T), GABAHARMASS (T Mensa). The fruit, dark red-brown, is ½ to ¾ in. diam.

It is possible that one or both of the recipes below apply to *Z. mucronata*.

Recipes

(a) Take YE GESHO FERE ('seeds' of GESHO) and use the juice for yellow.
(b) Fresh GESHO FERE are picked, crushed and mixed with water to dye cloth. The colour is black and fast.

Sources: Alemu Wolde Medhin (Arussi) gives recipe (a).
Arega Gebre (Kaffa) gives recipe (b).

Rumex abyssinicus - MEKMEKO (A), MOKMOKO, *MEKMOKA*, *MEKIMOKO* (A, T), CHOLDIA, DANGANO (Wollamo), *?DANGAGO* (G).

A tisane from MEKMOKO is used for whooping cough and the powdered root - mixed with butter - for skin diseases. Cufodontis adds that a laxative is obtained from the roots.

Note: Cufodontis gives *Chenopodium murale* as MOCMOCO, MOQMOQO in Acrur.

Historical notes: Bruce, J., in *Travels to Discover the Source of the Nile*, Vol. VI, pp. 399-400, describes 'Moc-moco', bruised, being used by the Agaws at Geesh to preserve butter, adding 'Brides paint their feet likewise from the ancle downwards, as also their nails and palms of their hands, with this drug.'
Valentia, G., in *Voyages and Travels*, Vol. III, p. 162: 'Coarse carpets are made in Samen, and Gondar, from the wool and hair of the sheep and goats ... They also procure a yellow dye from the Mocmoco ...'
Harris, W. C., in *The Highlands of Aethiopia*, Vol. II, Appendix, p. 409: 'Rumex arifolius (*Makmako*), frequent in swampy meadows, yields, in its fleshy root, a reddish dye for colouring butter.'
Von Heuglin, M. Tb., in *Reise nach Abessinien*, p. 328: '"Mogmogo": die Wurzel von *Rumex habessinicus*, zum Gelhfärben der Butter.'
Griaule, M., in *Le Livre de Recettes d'un Dabtara Abyssin*: '... on l'emploie également dans la fabrication de la couleur jaune: à cet effet, on le fait bouillir avec de la terre jaune et on décante.'

Recipes

(a) Take the root and hammer it with a stone. Mix with water. Fast colour.
(b) Chop the root and cook it with butter and water to obtain a rose colour.
(c) To 1 litre of water add a handful of chopped or powdered MEKMOKO and 30 grams (1 tablespoon) of alum and the same of salt. Boil at 100 - 120°C for two or three hours, then allow the wool to lie in the cooling water for at least twenty-four hours. Chemically treated water needs more colouring agent.
(d) Chop and boil the root of MEKMOKO and add some oil and lemon juice. Put the dye in a flat *mitad* (metal dish) and lay the grass to be coloured for basket-making in the liquid. Leave for twenty-four hours, away from sunlight.
(e) Chopped MEKMOKO and ABALO (see *Terminalia glaucescens*) are boiled together with *mashega* (parchment glue) as a dye for yellow clothes and/or *mesob* (basket) grass. Some people omit the *mashega* when dyeing clothes.
(f) The recipe of Griaule, given above.

Sources: Arega Gebre (Kaffa), gives recipe (a).
Taddesse Woldemeskel gives recipe (b), adding that the colour is fast.
Sig. Gilardi and *Ato* Tekle Haile of the Empress Menen Handicraft School Dye Shop, provided recipe (c), stating that the colour is 100 per cent fast if alum is used as a mordant. Using this same recipe, but rinsing afterwards in a hot soap bath produces a rosier, almost orange colour.
Wo Turuwork Teffera gives recipe (d).
Wo Tsehai Haile gives recipe (e).
Worku Ambie (Gojjam) simply says 'use dry' (as a crayon?).
Gebre Igziabhier Haile Mariam (Wollo and Addis Ababa) uses the spelling MEKIMOKO and added that 'the clothes of the monks of Debra Libanos or Waldebba are washed with Mekimoko'.
Tesfaye Fodone (Illubabor).
Wondimagegn Tefera (Illubabor) gives DANGAGO (G).
Mulat Admassu.
Endale Taddesse (Illubabor).

Rumex steudelii - TULT, TUTT (A), SHOMBORA, SHOMOBATA (T), ARAB-SARI (Harrar).

Recipes

(a) Squeeze the fruit of TULT.
(b) Use the root of TULT.

Sources: Million Abebe (Hararghie) gives recipe (a).
An old man employed by the School of Fine Art, Addis Ababa, gives recipe (b).

Solanum spp. - EMBWAI, OMBAI, OMBOI (A).

Historical notes: Cecchi, Antonio, in *Da Zeila alla frontiere del Caffa* (1886), Vol. I, p. 293: 'Il succo dei limoni e il solo mordente impiegato. Per tingerle en giallo, adoptera la scorza dell'albero "omboje" e rammollite coll'olio che estraggone dal frutto del "gurli": cosi preparate, esse servono di vestimenta ai pellegrini e ai monaci del paese.'

Luigi Amedeo di Savoia-Aosta, *La esplorazione dello Uabi-Uebi Scebeli,* p. 432, Solanum marginatum: 'I frutti di questa pianta sono usati comunemente in tutta l'Abissinia nella concia delle pelli, e come sapone per lavare la biancheria.' (The fruit of this plant is commonly used in all Ethiopia for tanning hides, and as laundry soap.)

Pp. 405-6, Cucumis laevigatus: 'Lesemplare à accompagnato dalla sequente nota del raccoglitore: "L'infuso della radice usasi per curare la blenorragia e anche per la fermentazione del *tec* (bevanda alcoolica), il frutto è usato come recostituente pei bovini esauriti. Nome indigeno *Umbai.*" '(The sample is accompanied by the following note of the collector. An infusion of the roots is used as a cure for gonorrhoea and also for the fermentation of *tej* (an alcoholic beverage), the fruit as a cattle tonic. Indigenous name UMBAI.)

Notes: With reference to the fruit 'gurli' mentioned by Cecchi, Dr Tewolde Berhan, Botany Department, H.S.I.U., says that the fruit of *Solanum spp.* is known by the name of ENGULI (T).

Lemordant quotes Guidi as saying that there are three types of *Solanum spp.* differentiated by name:

gäbär əmbway, the largest
zärç əmbway, the smallest
yä mədər əmbway, the medium.

He mentions that the acid juice is used to bleach cloth, quoting Dillon as saying that it is employed for tanning.

Recipes
(a) The yellow fruit of IMBUWAI will dye hands red if it touches them.
(b) The fruit of GABAR EMBEWAY is added to Black Ink to keep away flies.

Sources: *W/t* Azeb Desta (Harrar) gives the note (a).
Marigeta Tsege Tesfa Hunezu (Gondar) gives the note (b).

It is also used in the Black Inks of *Abba* Wubu Kidane Mariam and *Memhere* Hailu Man Wasinot.

Terminalia brownii - HUEBA, *WEIBO* (A), GUEVA, SHAHATT, VAIBA, WEIBA, WOIBA (T).

Cufodontis gives HOUEBA (?), GUERSÀ, UAIBA, UEIBÀ, UEBA, VAIBA, WEIBA, WOIWA (T), SCHAHATT, SCHEBATT, SSAHATT etc. (Tigre), and notes that the bark is used to give yellow.

Lemordant states that the tanning bark is used in decoction for jaundice and the stems for inhalations.

Note: The word *weiba* can be used to describe an 'orange' colour. There is often confusion to the European ear between V and B. It should be noted that for *Combretum guenzii*, Breitenbach gives WEIEVA (A), the flowers of which are yellow, often tinged with red; the fruit, 1-2 cm. long, is yellowish. In Ethiopia the family *Combretaceae* comprises the genera *Combretum*, *Terminalia* and *Anogeissus*.

Historical notes: Théophile Lefebvre and others in *Voyage en Abyssinie*, Vol. III, p. 241, writing on leather: '... on teint en jaune à la même manière (as red with KERATE) avec l'écorce nommée ouéba.' P. 244, on dyeing: 'Ils donnent au coton une couleur jaune paille avec l'écorce du ouéba, qui leur sert aussi pour les cuirs.'
Pearce, N., in *Life and Adventures in Abyssinia*, Vol. I, p. 196: 'The true mourning suit of the people of rank is a new white cloth, first dyed yellow with *waver*, the wood of a tree, which the monks use to dye their garments.'
Girard, in *Souvenir d'un Voyage en Abyssinie*, pp. 241-2: '... ils donnent au coton une couleur jaune avec l'ouéba, écorce d'arbre du pays.'
Wylde, A. B., in *Modern Abyssinia*, p. 355: '...and a very pretty tree called by the Abyssinians the Waiva, which grows to a large size... end of each branchlet a bunch of purple-coloured flowers which contain a flat seed about the size of a shilling, that is used by the priests to dye their garments a yellow colour, a lighter shade than gamboge, and the dresses of the buddhist priests in Ceylon and those worn by the Abyssinian monks are nearly of the same tint.' It seems probable that Wylde misread his notes as the flowers of *Terminalia brownii* are white and foul-smelling. The most likely 'purple-coloured flowers' would seem to be those of *Millettia ferruginea*, which, according to Breitenbach, has lilac-coloured flowers borne in clusters. The seeds, in pods, are dark brown, flat, and 2 cm. in diameter. When pulverized they are used as a fish poison.

Recipe: Take YE WOIBA LIT (the bark of WOIBA). The inner bark should be used. Crush it and soak in a container with water for 24 hours or longer. Boil for a long time and then strain. A little gum and/or salt may be added to the dye bath.

Sources: Mengesha Yibrah (Begemeder) writes WEIBO (A), mentioning that priests' robes and turbans and women's mourning dresses are dyed with this bark.
Hadgu Bariagaber and Ogbai Zeru both give a slightly different recipe: the dried bark, chopped up, is soaked in water for about three days. 'Monks dye their clothes in this liquid.'
Ato Alemayehu Moges (Gondar and Addis Ababa) adds that a little gum, preferably that of *Acacia spp.*, and/or salt may be added to the simmering dye bath. He thinks it possible that pieces of the roots may also be included. *Ato* Alemayehu remembers as a child helping with other children to collect the WAIBA; clothes were dyed in this mixture by Government order during the Italian invasion.
W/t Azeb Desta (Harrar) gives the name WOIBA as used round Harrar.
Bissrat Fesahaye (Tigre) says the colour is fast.
Assefa Asghedom (Kaffa).
Aregawi Woldemariam (Eritrea).
Arega Gebre (Kaffa) and
Yacob Kassahun also say that it is a fast colour.
Yehdego Melles (Tigre).
Arefa Gebresadik (Tigre).
Worede Isaac (Eritrea).
Aleqa Gebre Selassie (Selale).
Aleqa Wolde Medhin (Menz).
Ato Imeshaw Temtin (Bulga, Shoa) gives WEIBA and adds 'not the one for oil'. Possible this is a confusion with *Olea spp.* WEIRA.

Terminalia glaucescens - ABALO, ABELLO (A), TAGGIEH (A Gondar).

Note: It is probable that the recipes given below for ABALO refer to *Terminalia glaucescens*, but the name ABALO is one of those which refer to more than one plant (owing to medicinal connotations). Lemordant describes ABALO as a greenish-yellow field plant, identical to WAGINOS. (Mooney gives *Brucea antidysenterica* as WAGINOS (A), ABALO (G).) Griaule says that it is the plant Reinish calls ABALUW. Schweinfurth gives AWALO as *Premna resinosa*.

Griaule: '... plante dont on se sert pour parfumer le vase contennant du lait.' Field plant, yellow-green, symbol of plenty. One is inclined to agree with Strelcyn, who states that the identity of ABALO is uncertain, the term probably indicating different plants, among them *Terminalia glaucescens*, *Crotalaria lachnocarpoides* - for which Mooney gives ABU-ALU (A), OKOKO (K) - *Brucea antidysenterica*, and *Combretum molle*, for which Cufodontis gives AVALU (A), ABELOA, AMFERFARO, ANFERFARO (Tigre & Ghinda), HATHIBA, HAZIBA (T Hamasen), SESSOI, SSOSSOI, SSOSSOUE (T Acrur) and WEIBA (Cherendai).

Recipes

(a) Take the bark of the tree ABALO. Soak for twenty-four hours in water, and then dip in the cloth. Gives a fast dark yellow.

(b) The bark of ABALO is cut, crushed with a pestle and mortar and water added. The colour is red. (Probably the mixture should be boiled.)

(c) The bark, leaves and buds of ABALO are boiled together for yellow for painting. For dyeing, both yellow and red may be obtained (possibly by using the bark alone for red?).

Sources: Taddessa Woldemeskel gives recipe (a).
Mekuria Degu (Shoa) gives recipe (b), noting the colour as red.
Aleqa Wolde Medhin (Menz) gives recipe (c), but may be referring to two different plants of the same name.

?Nuxia congesta - CHOJO (?A).

It was suggested by Dr Amare Getahun, Alemaya, Haile Sellassie I University, that from the name alone this might be *Nuxia congesta*, given by Mooney (among other names) as CHIACHIOA and JAJO (A), and by Cufodontis as CIACIOÀ or JAAHJOH (A). Mooney also gives *Otostegia integrifolia* as TSCHO-TSCHO (A); and *Premna schimperi* as TSCHO-TSCHO (A). For *Otostegia integrifolia* Cufodontis gives TSCHOTSCHO (A), adding that an inhalation of the roots relieves respiratory diseases. Lemordant describes č̣oč̣o as a tree of the *qolla*, quoting Baetemann as suggesting that it is *Pterolobium abyssinicum*.

Historical note: Harris, W. C., in *The Highlands of Aethiopia*, Vol. II, Appendix, p. 407: 'Scrophularia frutescens (*Djodjo*), with a strong smell of camphor, is used as a febrifuge and charm.'

Recipe: Crush and boil the leaves of CHOJO for a fast yellow colour.

Source: 1968 notes, name lost.

Berberis sp.

Historical note: Harris, W. C., *The Highlands of Aethiopia* (1844), Vol. II, Appendix, p. 412. '*Berberis tinctoria* of the forests yields a good yellow dye for mourning apparel.'

Burger, W. C., *Families of Flowering Plants in Ethiopia*, 1967, notes: '*Berberis holstii* is the only species of this family native to eastern Africa'. Mooney gives *Berberis holstii* as ZINKILLA (A).

Note: The only note of yellow being used today for mourning purposes occurs under *Terminalia brownii*. No recipes for *Berberis sp.* have been collected by the writer.

ORANGE

Mimusops kummel	KUMEL, KUMAL(T)
Oxalis anthelmintica	HABOCHEGO (T)

Mimusops kummel - ISHEH, ISKYEH, SHEH, SHI (A), KUMMEL (A, T), BURURI, KOLADI, KOLATTI (G), NEGA-KADADI (G-H), DEMBI, MITTO (G-Betscho), GAYU (K).

Breitenbach describes the fruit thus: 'Fruit ellipsoid, slightly more than 2 cm. long, yellowish when ripe, edible, containing a single stone-seed'.

Breitenbach also lists *Sclerocarya birrea*: KUMAL, GOMALES (A), ABENGUL (T) and describes the fruit as 'pale yellow, plumlike, 3-4 cm. in diameter with a tough skin and juicy mucilaginous flesh which it is difficult to separate from the stone, edible'.

Recipe: Squeezed KUMEL (T) and water gives a fast orange colour.

Sources: Yacob Kassahun gives the above recipe.
Tsegahun Mebrahtu (Tigre) gives the colour as 'dark pink': he later spoke of it as 'violet', pronouncing the name as KUMAL and suggesting that it was a 'spice'.

Oxalis anthelmintica - ABBA-TSHEGO, HABBE-TSCHOKKO, MITSHA-MITCHO, *HABOCHEGO* (T).

Cufodontis gives HABBE-TSCHAKKO (T) and MITCHAMITCHO, MITCHIÁ-MITSCIÓ (A) and says the tubers are used medicinally.

Recipe: Squeeze HABOCHEGO (T) and add water for light orange.

Source: Yacob Kassahun.

RED

Aloe sp. possibly *A. steudneri*	SETE RET, SET ERET (A)
Capparis sp.	ANDEL (T), GUMERO (A)
Carthamus tinctorius	SUF (A)
Cassia singueana	HEMBO-HEMBA, BUSHA (T)
Cicer arietinum	SHIMBERA (A)
Crassula ellenbeckiana	LASHALASHO (Harrar)
Crinum sp. or *Urginea sp.*	YEDJIB-SHINKURT (A)
Duranta repens	MWATISH (A)
Ensete ventricosum	ENSET (A)
Eucalyptus globulus	BAHARZAF (A, G)
Euphorbia candelabrum or *Opuntia sp.*	KULKWAL (A)
Ficus sp.	WARKA (A), KILTU (G)
Flemingia grahamiana	WARS (Ar)
Hypericum sp.	AMECHA (A)
Impatiens spp.	ENSOSILA, GIRSHIT (A), ELAM (T)
Indigofera arrecta	?ILAM, ILAM WEKARIA (T)
Kalanchoe sp.	YEWUSHA ABEBE (A)
Lawsonia inermis	HENNA (Ar), ELAM (T)
Osyris abyssinica	KERET (A), WATO (G)
Pterolobium stellatum	KANTUFFA (A) See also Black Inks
Rhamphicarpus sp.	YESET LEB (A) See Red Ink
Rosa spp.	KEGGA, TSEGYE-REDA (A)
Rubia discolor	ENCHIBIR (A), BARTUTA (Kam)
Rubus spp.	INJORRE (A)
Rumex nervosus	EMBEWACHO (A) See Red Ink
Saccharum officinarum	TINKISH (A)
Schefflera sp.	ADIM (T)
Solanum nigrum	TIKUR-AWITT (A)
Ximenia americana	MELLAU (T), INKOI (A)
Unidentified	AGOR GOBASH (A) See Red Ink
Unidentified	EJEAKELA (A)
Unidentified	ETAIBETER (Gurag)
Unidentified, *?Pavonia sp.*	TOHGO (G)
Historically noted:	
Acacia bussei	GALOL (Som)

Cordeauxia edulis	GUT (Som)
Discopodium penninervium	AMORAROO (A)
Unidentified	HADDIE

RED INK RECIPES

Aloe sp. possibly ***steudneri*** - SETE RET (A, T).

Mooney gives most *Aloe spp.* ERRET (A).

SETE RET identified by name alone by Mr M. Gilbert, the Herbarium, H.S.I.U., who notes that SET ERET (women's aloe or female aloe) is used for certain highland Aloes which have more or less smooth margins to their leaves. The sap, at first clear, turns deep purple on exposure to air. This name probably covers a number of species such as *A. steudneri* and *A. percrassa*. Species with distinct marginal teeth are known as WEND ERET (Man's aloe or male aloe).

Recipes

(a) Chop up SETE RET and stir in water. When white cloth is boiled in this mixture it will turn red.

(b) Cook it like 'a cabbage' and then pound it to extract a kind of greasy jelly. Filter it. Can be used as a painting medium; though not very satisfactory it gives a good gloss. It may also be used as a glue.

Sources: Mekuria Degu (Shoa) gives recipe (a).
Aleqa Gebre Medhin (Menz) gives recipe (b) adding that other types of RET may be used but that SETE RET is the best.

Capparis sp. - ANDEL (T), GUMERO (A).

Capparis micrantha - ANDAL, ANDEL, ANDELO (A, T), ARAN GAMMA, GURACHIA (G).

Capparis tomentosa - GIMERO, GUMARU, GUMURO (A), ANDAL, ANDEL, ANDELO (A, T), GOMBOR (Harrar).

Recipes

(a) The juice of the ripe fruit of ANDEL (T) gives red.

(b) The fruit is boiled together with the roots of GUMERO to make red.

(c) The roots of GUMERO are boiled together with the bark of INKOI (*Ximenia americana*) for a rose red.

Sources: Hadgu Bariagaber and Ogbai Zeru give recipe (a).
Aleqa Gebre Selassie (Selale) and *Aleqa* Wolde Medhin (Menz) give recipe (b).

Ato Imeshaw Temtin (Bulga, Shoa) gives recipe (c).

Carthamus tinctorius - Saf flower. SUF (A, T).

Cufodontis gives CHOUF, SCHUF, SCHUHF, SCIUFF, SSUF, SUF, YAHAYA-SUF (A, T), SUFI (G), grows wild and is also cultivated for its oil and yellow pigment. (A well-known dye prepared from the flower heads of the safflower is used in the Orient for dyeing silk and cotton light red.)

As well as *Carthamus tinctorius*, *Helianthus annuus* is also known as SUF.

Historical notes: Harris, R., in *The Highlands of Aethiopia*, Vol. II, Appendix, p. 409: 'Carthamus tinctorius (*Suf*), extensively cultivated in Efát for the oil of the seeds and for the dye yielded by the flowers.'

Von Heuglin, M. Th., in *Reise nach Abessinien*, 1874, p. 326, notes it as cultivated and in use: 'Cathamus (šüf) von Tiefland bis über 7000)'. On p. 183 it is referred to as 'Saflor'.

Wylde, A. B., in *Modern Abyssinia*, 1901, p. 267: 'The safflower is grown in some parts of the country, and other dyes are found growing wild; these plants are preserved and not cut down when clearing the ground for cultivating', and p. 243: ' ... the beautiful souf, a thistle-like plant with bright orange and red flowers bearing a white seed'.

Mérab, Dr P., in *Impressions d'Ethiopie*, 1929, Vol. III, p. 204: '... voici les plantes les plus utilisées par les Ethiopiens en teinturie: ... 3ème les fleurs du "souf" (*Carthamus teinctorius*) qui n'est autre que la variété ethiopienne de notre safran'.

Recipes

(a) Pounded SUF flowers give red.

(b) SUF is boiled for yellow. Sometimes butter or the fat from animal bones is added, but this tends to make the dyeing uneven. In order to ensure an even colour most dyers prefer not to use fat. The colour is fast.

Sources: *W/t* Azeb Desta (Harrar) gives recipe (a).
Alem Teklu (Tigre) gives recipe (a).
Ato Imeshaw Temtin (Bulga, Shoa) gives recipe (b).

Cassia singueana - BISHBISHA (A), AMBA-AMBO, *HEMBO-HEMBA*, *HAMBO-HAMBO*, *SAMBO-HAMBO*, *BUSHA*, BUS HOMBOI (T).

Cufodontis gives: BIBISHA, BOSCBOSCIÁ (A), MBOI (Tigre), SSAMBO-HAMBO (Saha & T), BUHSS, BÛSS (T Asmara), AMBA-AMBÓ, AMBA-HAMBO, AMBÓ-AMBÓ, HAMBA-HAMBO, HAMBE-HAMBE, HAMBE-HAMBO (T Mensa & Tigre), HOMBOI, HUMBOI (T Hamasen).

Note: Breitenbach gives BIRBISSA and BIRBIKSO (G) for *Podocarpus spp.* for which Mooney gives BIRBIRSA, BIRBISSA (G-W). The fruits of these trees produce an oily fruit juice used for medicinal purposes. In *Common Poisonous Plants of East Africa,* Bernard Verdcourt and E. C. Trump, 1969, we note the following local names for *Cassia singueana*: HUMBAHUMBA (Vidunda), MAHUMBAHUMBA (Kisagara), MUHOMBAHOMBO (Kihehe).

Strelcyn writes: '... il y a une racine qui s'appelle hambö en langue Tigrinia. Il s'agit, selon toute probabilité de la racine de Hambo hambo, arbre dont la racine sert à teindre en rouge le tour des yeux'. Coulbeaux-Schreiber 12.

Historical notes: Lefebvre, in *Voyage en Abyssinie,* 1845, Vol. III, p. 241: '... cuir rouge, teint avec l'écorce de l'arbre appelé "Kerate", qu'on mêle a la fleur séchée bebetcha'.

Cecchi, in *Da Zeila alla frontiere del Caffa,* 1886, Vol. I, P. 293, 'Per colorare le pelli in rosso, il "faki" si serve della scorza dell'albero chiamato "Kerate", mescalata ai fiori secchi de un'altra pianta chiamata bebeteja'.

Von Heuglin, M. Tb., in *Reise nach Abessinien,* 1874, p. 328: 'Cassia sp.? ein Strauch, gibt geibrote Farbe'.

Recipes

(a) The root of ADIM (T) (*Schefflera sp.*) and the root of the HEMBO-HEMBA (T) trees when soaked produce red.

(b) The dried leaves and bark of KERET (*Osyris abyssinica*) mixed with the dried leaves and bark of BUSHA, together with water, are used as a dye.

Sources: Yehdego Melles (Tigre) gives recipe (a).
Ato Gelahun Abate (Eritrea) gives recipe (b) and the name BUSHA.
In 1968 notes - source and recipe lost - 'BISHBESHA 8 days' was entered by the writer under the colour yellow.
Hadgu Bariagaber (Eritrea) gives the spelling SAMBO-HAMBO.
Various other pronunciations were noted in discussion.

Cicer arietinum - SHIHU, SHIMBERA, SHUMBRA (A), ATER-KHIJEH (T).

(The dried chick peas, pounded, are used as a cosmetic wash by ladies, as also are dried *Faba bona* BEKELA (A, G).)

Recipe: Pounded YESHIMBIRA ABEBE (flowers of SHIMBERA) give a pink colour.

Source: *W/t* Azeb Desta (Harrar).

Crassula ellenbeckiana - *LASHALASHO* (Harrar).

Specimen identified by Mr M. Gilbert, the Herbarium, H.S.I.U.

Recipe: The buds of LASHALASHO, before they are opened, are pounded with a little water. The juice is used to dye white cloth pink.

Source: *W/t* Azeb Desta (Harrar).

Crinum sp. or ***Urginea sp.*** — YEDJIB SHINKURT (A).

Recipe: First pound the root of YEDJIB SHINKURT and then boil. This gives a transparent red colour, almost a rose-red, as in the various 'lakes'. Used for painting.

Source: *Aleqa* Wolde Medhin (Menz).

Duranta repens - *MWATISH, MWAT'ESH* (A).

Identified by Dr Tewolde Berhan, the Herbarium, H.S.I.U., from the name alone.

Described by the source below as a 'creeping plant', and by *Ato* Alemayehu Mogus (Gondar and Addis Ababa) - who gives the second pronunciation - as a parasitic plant growing over bushes.

Recipe: Chop the leaves of MWATISH (A) and boil them. Filter after three days. Fast colour.

Source: Taddesse Woldemeskel.

Ensete ventricosum - ENSET, INSET, GUNA-GUNA (A), KOCHO, KOTCHO, WERKE (G), KOBA (G-Bale), KOICHO (K).

Recipe: The flower of the false banana called *SHIRA*, 'obtained only from old plants in the Southern Provinces'. The flowers grow in bunches. These bunches should be cut and soaked for some hours in cold water.

Source: Dendir Dansamo (Sidamo and Shoa).

Eucalyptus globulus - BAHARZAF (A, G), BAHIRZAF, BARZAF (A, T), AKACHITTA, ATAKILI (G), *KALAMETUS* (T).

Recipe: Eucalyptus tree gum, called ENDADIE, gives a red colour.

Source: Alem Teklu (Tigre) who also gave KALAMETUS.

Euphorbia candelabrum - KALKWOL, KALQUOL, KINCHAB, KULKWAL, KULQUAL (A, ?T), ADAMI, HADAMI (G), AKIRSA (G-W), KATSCHO (G-K).

Opuntia spp. - KULKWAL, KULQUOL (A).

Note: Both the above species have red fruit.

Historical note: Mérab, P., in *Impressions d'Ethiopie*, 1929, describes QULQUAL as being used in tanning and also for rendering baskets water-tight.

Recipe: The red fruit at the top of KULKWAL (A) can be crushed and prepared to give a fast colour.

Source: Debebe Haile Giorgis (Harrar).

Ficus (probably) ***vasta*** - SHUOLA, WARKA (A), WORRKA (A, G), DAHRO, DARO (T), DEMBI, KILTI (G).

Ficus brachypoda - KILITU, KILTU (Gudgi).

Mr M. Gilbert, the Herbarium, H.S.I.U., says the name SHUOLA (A) is generally used for *F. sycamorus*.

Recipes

(a) The smooth inner bark of WARKA is collected and pounded into fine fibre. Soak in water. After a week or so add more water and stir well. Filter out the fibre. Fast colour.

(b) The 'fruits' of WARKA are gathered and roasted. When roasted dark they are taken out and ground. The powder is mixed with water and used for writing. It is a fast colour.

(c) The root of KILTU, if rubbed on anything when dry, changes to red.

(d) The branches of WORKA were boiled to produce a 'red' dye for camouflaging clothing during the Italian occupation.

(e) The WARKA bark is boiled with salt for red.

Sources: Bayelegne Meshesha (Bale) gives recipe (a), adding: 'Used for dyeing, painting, and decorating houses'.

W/t Azeb Desta (Harrar) gives recipe (a), using the name SHOLA.

Taddesse Woldemeskel gives recipe (b).

W/t Azeb Desta gives recipe (c), listing KILTU as distinct from SHOLA, which she gives for recipe (a).

Ato Gebre Tsadik Wolde Meskel gives the information in recipe (d).

Ato Imeshaw Temtin (Bulga, Shoa) gives recipe (e).

Flemingia grahamiana ≡ F. rhodocarpa, Moghania rhodocarpa, Eriosema erytbrocarpa and **E. robustum** — WARAS, WARUS, WARS, WURS (Ar), WARSI (Harrar).

Historical notes: Major F. M. Hunter and Lieut J. D. Fullerton, in *Reports on Somali Land and the Harrar Province*, pp. 98-9, state:

> 'Wars is not now raised in the neighbourhood of Harrar from seed sown artificially, and it is left to nature to propagate the shrub in the surrounding terraced gardens. The plant springs up among jowari, coffee, etc., in bushes scattered about at intervals of several yards more or less. When sown, as among the Gallas, it is planted before the rains in March. If the soil be fairly good, a bush appears in about a year. After the berries (pods) have been plucked, the shrub is cut down to within six inches of the ground. It springs up again after rain, and bears a second time in about six months. This process is repeated every second year until the tree dies. Rain destroys the berry (pod) for commercial purposes. It is therefore only gathered in the dry season ending about the middle of March. The bush grows to a maximum height of six feet, and it branches off close to the ground. The growth is open and the foliage is sparse. Each owner has a few acres of land. In the end of February 1884, the following processes were observed: The leaves (?fruiting shoots) of some plants were plucked and allowed to dry in the sun for three or four days. The picking is not done carefully, and a considerable quantity of the surrounding twigs, etc., is mixed with the berries (pods). The collected mass was placed on a skin heaped about six or eight inches high, and was tapped gently with a short stick about ½ in. thick. After some time the pods were denuded of their outer covering of red powder, which fell through the mass on to the skin. The upper portion of the heap was then cleared away and residual reddish-green powder was placed in a flat woven grass dish with a sloping rim of about an inch high. The receptacle was agitated gently and occasionally tapped with the fingers, the result being the subsidence of the red powder, and the rising to the surface of the chaffy refuse, which latter was carefully worked aside to the edge of the dish and removed by hand. This winnowing was continued until little remained but the red powder. (No great pains are even taken to eliminate all foreign matter.) A roll was sold in 1884 for about thirteen piastres - 1 Re. 10 As. nearly. Wars is sent to Arabia chiefly to Yemen and Hadhramaut, where it is used as a dye, a cosmetic, and a specific against cold. In order to use it, a small portion of the powder is placed in one palm and moistened with water, the hands are then rubbed smartly together producing a lather of a bright gamboge colour which is applied as required.'

In Appendix C of the same book, Lists of Imports and Exports, Wars is noted as imported from the Interior (to Harrar) and exported to Zaila, Berbera and Aden: imported by the five gates and exported from the gates Bab al Rahmah and Fatah.

Sir Richard Burton, in *First Footsteps in East Africa*, Vol. II, p. 27, confuses Wars with Safflower (*Carthamus tinctorius*). 'The Wars or Safflower is cultivated in considerable quantities around the city of Harrar: an abundance is grown in the lands of the Gallas ... This article, together with slaves, forms the staple commerce between Berbera and Muskat. In Arabia, men dye with it their cotton shirts, women and children use it to stain the skin a bright yellow: besides the purpose of a cosmetic, it also serves as a preservative against cold. When Wars is cheap at Harrar, a pound may be bought for a quarter of a dollar.'

Yusuf Ahmed in *An Inquiry into Some Aspects of the Economy of Harrar*, 1825-75 quotes Major Mohammed Moktar (Muhtar Bey) in the *Bulletin de la Société Géographie du Caire*, 1876, 'Notes sur le pays de Harrar'. Moktar remarked that among plants cultivated for condiments he noticed Safflower (wərsi) ... Moktar called it warth (p. 16). Either Muhtar Bey was quoting from Burton, or there is a possibility that *Carthamus tinctorius* is, or was, known locally in Harrar by the name of WARSI.

Recipe: The 'flower' of WARSI, dried and pounded, gives yellow. If mixed with water a piece of cloth soaked in it will turn red.

Source: *W/t* Azeb Desta (Harrar), who gives another recipe for SUF flowers (*Carthamus tinctorius*).

Hypericum sp. (names are the same for *H. revolutum* and *H. quartinianum*) - AMEDJA, AMECHA, AMEJA, AMIDSCHA (A), GORGORO, HINNEH, INNI (G), MITO (G-H), GARAMBA (G Bale), EDERA, GARARU-BICHU (G-Sid), GRAM-BICHU, MUKA-AFONA (G-W), MUKOFONI (G-K), AWETSCHA (T).

Recipe: Both the flowers, which are yellow, and the roots, are boiled together until most of the water has evaporated, to provide red for ink or painting.

Source: *Aleqa* Gebre Selassie (Selale), who gives the name AMECHA or AMEDJA.

Impatiens spp. and possibly *Conyza sp.*

For *Impatiens tinctoria* Cufodontis gives: EŠOŠILA, GESCHEIOAHT, GHIRSCHIT, GIRSHID, GURSCIT, GUSCHEREDD (A), ELLAM, ELLAME, ELLAMIE, ELLEN, ENSSESSELLA, ENSSOSELLA, HELLAM, SASSULLA, GURELILE, GURRELIL (T), HANÚ (Hamasen), UTSCHILLO (K), OSSILE (G), WOSOLUA (Wollamo).

S. Edwards gives *I. rothii* as GURSHIT (A), a species growing in well-drained areas of the plateau. She gives *I. tinctoria* as INSOSILLA (A, G-Illubabor), SASULLA (T).

Names collected in recipes below: EMSOSLA, ENSOSELA, INSHOSHELA, INSOSELLA, GURSHERED (A), ENSOSILA, INSSOSSULA (A, G), SOSOLA (G), ELAM, ILAM, GIRSHIT, GUSHIT, ENSESULA, INSOSILA, INSOSLA, SASELA, SASILA, SOSILA (T).

Historical notes: Harris, W. C., in *The Highlands of Aethiopia*, Vol. II, Appendix, p. 412: 'Some *Balsamineae* grow in shady places: one of them, Impatiens grandis (*Girshid*), has a tuberous root, with the juice of which the women paint their palms and faces red.' And in Vol. III of the above, Ch. XVIII, pp. 158-9: '... the lady of rank completes her toilet by dying her hands and feet red with the bulb called ensosela, ...'

Von Heuglin, M. Th., in *Reise nach Abessinien*, 1874, p. 328, 'Impatiens tinctoria (Elamie) zum Färben der Hände.'

Dr Mérab, in *Impressions d'Ethiopie*, 1929, Vol. III, p. 402: '... voici les plantes les plus utilisées par les Ethiopiens en teinturie: 1ère racine de "guerchet" (*Impatiens tinctoria*), la même que celle qui sert a teindre les mains en rouge ...'.

Note: There exists some confusion over local names. For example ELAM (T) is used for *Lawsonia inermis* (Henna), as well as for *Impatiens tinctoria* and most probably other species or varieties.

Lemordant gives: ənsosəlla and gəršət '... les deux termes ne sont pas équivalents. En réalité ənsosəlla désigne *Impatiens tinctoria*, Nob., et gəršət *Conyza abyssinica*, C.-H. Schultz. La confusion provient du fait que la seconde est utilisée comme succédané de la première car son prix est moins élevé.' (For medicinal use.)

Aleqa Gebre Selassie (Selale) and *Aleqa* Wolde Medhin (Menz), two elderly painters, who have used the colour in their work, maintain that the plant is called GURSHERED and the colour INSOSILA.

Mengesha Yibrah (Begemeder) describes ELAM (T) as a 'colour-giving root plant used for beautifying the hands and legs, mostly by Tigrai ladies' and 'SASILA, the same as above'.

Tsegahun Mebrahtu (Tigre) gives the names GUSHIT and ELAM, describing them as 'similar in appearance'.

Dendir Dansamo (Sidamo and Shoa) gives the names INSOSILA and GIRSHIT, and says '... both are herbs growing by riversides and in the garden; the roots are dug out ...'.

Ogbai Zeru and Hadgu Bariagaber (Eritrea), after a home visit, describe ILAM as roundish, usually about the size of a clenched fist, and SOSILA as the size and shape of a large carrot.

Wo Almaz Kebede describes GURSHERED as being of a lighter colour than INSOSILA, saying that when the former is cut and exposed to the light it turns yellowish, while the latter turns a deeper colour.

Ogba Michael Tekle supplies the following information:

> 'ILAM is carrot-like, and grows wild in the countryside. There are two kinds of ILAM, a large one growing in the highlands, and a slightly smaller one growing in the lowlands. Both of them grow in dense and moist places, under big trees. It is the lowland variety which is called INSOSLA in Amharic. The ILAM grows to a maximum height of one metre and is much in demand by both urban and rural married women. Girls use it to paint their palms and nails only, as to paint the feet is a sign that the girl is no longer a virgin. Young women use it a lot, and also prostitutes and widows, the aim being to attract men for prostitution and remarriage. ILAM is also used for the same purpose by *sewa* sellers, who are called *comoro*. They are makers of home-brewed beer. The young boys and girls search out mature plants with yellowing leaves, and dig up the fruit and wash it. It is then cut into pieces and left in a container for at least 12 hours. Finally and usually at evening, the chopped ILAM or INSOSLA, as it is called in some areas of Tigre, is heated on the fire in the container, and the women put their feet and hands into it and wash them with the hot chopped pieces of ILAM for about half an hour, and the girls do their palms and nails. Sometimes they wrap their hands overnight. It produces a dark reddish colour. However I have been told that the small ones which grow in the lowlands give an even darker colour, hence they are better.'

Recipes

(a) Remove the outside covering of the root and cut into pieces on a stone so as to loose no juice. Dry it in the sun and then grind it. Keep the powder in a bottle. Use stirred with water, though it should be left in the water for some time before use. There is some difference of opinion as to whether the colour is fast.

(b) After grinding the inside of the root, collect the semi-liquid substance and use it at once.

(c) Cut the root in pieces and expose to sunlight for a day. Add water and boil. Then ladies can put their hands and feet in the container.

(d) For dyeing cloth. Salt and oil should be added to the water in which the roots are boiling.

(e) Crush the root, then put it in leaves and add some water. Tie to the hands with string for 6-8 hours.

Sources: Gebre Behute (Gojjam) gives recipe (a) and spelling ENSOSELA.

Dendir Dansamo (Sidamo and Shoa) gives recipe (a) and spellings INSOSILA and GIRSHIT.

Lemma Areru (Gemo Gofa) gives recipe (a) and spelling ENSOSILA.

Wondimagegn Tefera (Illubabor) gives recipe (a) and spelling INSOSILA.

Million Abebe (Hararghie) gives recipe (a) and spelling INSOSILA.

Aregawi Woldemariam (Eritrea) gives recipe (b) and spelling INSOSILA.

Bissrat Fesahaye (Tigre) gives recipe (c) and spellings SASELA and ELAM, adding: 'It is like a carrot, an underground root.'

Arefa Gebresadik (Tigre) gives recipe (c) and spelling GIRSHIT.

Tsegahun Mebrahtu (Tigre) gives recipe (c) and spellings GUSHIT and ELAM.

Gebre Igziabher Haile Mariam (Wollo and Addis Ababa) gives recipe (c) and spelling INSSOSSULA.

Chane Addis (Eritrea) gives recipe (c) and spelling ENSOSELA.

Wo Mulugojjam Assaye, who heard it from *Wo* Fikerte, gives recipe (d).

Tessema Ta'a (Wollega) gives recipe (e) and spelling SOSOLA (G)

Further variations in spelling:

Tesfaye Fodone (Illubabor) gives INSOSELLA.

Tesfaye Alemu (Gojjam and Shoa) gives EMSOSLA.

Gebre Yohannes (Tigre) gives ENSESULA.

Indigofera arrecta - ?ILAM, ?ELAM (T), *ELLAM BUKARIA*, *ILAM WEKARIA* (T).
(Wild or 'fox' ELAM)

Cufodontis gives *Indigofera arrecta*: DAGENIEG, DEGENDEG, DEGENIEG, DIGHENDIG, DIKINDIK (T), ELLAM-MOKHARIA (T-Acrur), SCIAFIHA and ADBACHE-AUSINA (?T), and names it the principal dye species in Africa. Cufodontis gives *Indigofera articulata*: ELLAM-HABUT, HELLAM-HABUT (Tigre-Mensa), HABBI HASEI (T), saying that it is a suitable species for preparing indigo. He notes having seen *I. tinctoria* but gives no local names for it. Also, *I. spp.* HENNÁ (G), used for dyeing.

Historical notes: Valentia, G., in *Voyages and Travels*, 1809, Vol. III, p. 162: '... coarse carpets ... which are dyed red and light blue; ... the latter from a plant resembling *Indigofera* ... they have no dark blue.'

Von Heuglin, M. Th., in *Reise nach Abessinien*, 1874, p. 328. 'Indigo wächst wild, wird jedoch nicht benutzt.'

No recipes using *Indigofera* for blue have been collected.

Note: The following recipes are given under *Indigofera arrecta*, but owing to the common confusion over names it is by no means certain that this is their correct place. As opposed to the wild or 'fox' ELAM, the 'true' ELAM may be either *Impatiens tinctoria* or *Lawsonia inermis*.

Concerning the ELAM, or ILAM (T) under reference, sources either describe the plant as growing in a colder climate than *Lawsonia inermis* or simply refer to it as a different plant. *Indigofera arrecta* may not be the correct identification.

Recipes

(a) The root of ELLAM BUKARIA (T) is used.
(b) The small leaves of ILAM WEKARIA (T) are cut into pieces.
(c) The leaves of ILAM, dried, ground and mixed with water, are used by women to dye hands and feet.
(d) The plant ELAM is dried and powdered.

Sources: Mengesha Yibrah (Begemeder) gives recipe (a) and the name ELLAM BUKARIA (T), describing it as a very 'fugitive' colour, mostly used by small children. As the root is used, in this case the 'true' ELLAM is evidently *Impatiens tinctoria*.

Ogbai Zeru (Eritrea) gives recipe (a) and the name ELLAM WEKERIA (T), the K being pronounced as H.

Ogba Michael Tekle gives recipe (b) and the name ILAM WEKARIA, describing it as a bush with very small leaves. 'It is not known in the urban areas.' He thinks ILAM WEKARIA may be another name for HINNA (*Lawsonia inermis*), and that if it were to be prepared in the same way would give a similar colour. It is mostly used by young girls in the countryside who use it to wrap their fingers and toe nails for about 15 minutes. When the wrapped chopped leaves are removed there is left a 'violent' colour.

Idris Osman Samra (Eritrea) gives recipe (c) and the name ILAM (T) and says: 'The plants grow in a cold climate, especially the Highlands of Eritrea. It is exactly like HENNA (*Lawsonia inermis*) which grows in a hot climate.'

Abdulkadir Ahmed (Eritrea) gives recipe (d) and the name ELAM (T). Like the above-named, he also describes it not as HENNA (T), (*Lawsonia inermis*), but another plant, and Ogba Michael Tekle describes HINNA (*Lawsonia inermis*) as being similar to ILAM, having the same use, but as being more expensive. Woreda Isaac (Eritrea) was told by two separate Eritrean ladies that the leaves of 'i:lam' (sic) were used for dark red, as also were the stems of 'hi:na' (sic) (*Lawsonia inermis*).

Kalanchoe sp. possibly ***densiflora*** - *YEWUSHA ABEBE* (A).
Described as 'a plant found round Harrar town'.

The above identification by Mr M. Gilbert, The Herbarium, H.S.I.U., was made from a specimen of YEWUSHA ABEBE with yellow flowers. It is however possible that the recipe below may appertain to another plant.

Lemordant states that the names of many unidentified plants refer to dogs, thus starting *yä wuša* ... and Griaule mentions a *yewusha dinitch* as being unidentified, though Strelcyn gives 'dinich du chien' as *Coleus edulis*, the Galla potato. According to Dr Tewolde Berhan, The Herbarium, H.S.I.U., one species of *Indigofera* goes under the name of YEWUSHA DIGITTA.

It might be possible that in this case WUSHA is a corruption of WUSHISH. In Almeida (Beckingham and Huntingford), *Some Records of Ethiopia 1580-1645*, 1954, there is a reference to 'vegetables like turnips that they call uxixes and daniches. It bears a flower like saffron, and like it, this too dyes vermilion and yellow'. The footnote states: 'Wušiš and dennič, edible roots both of which have been compared with the potato. Bruce (Vol. VI, Ch. 12) identifies the 'denitch' with the Jerusalem artichoke'.

Mooney gives WUSHISH (T) as being both *Cucurbita pepo* and *Coccinea abyssinica*. Lemordant gives the latter as the probable identification of wušəš describing it as a plant of the Simien.

Recipe: When squeezed, the juice of YEWUSHA ABEBE gives a fast colour 'which can be used for many purposes'.

Source: Debebe Hailegiorgis (Harrar).

Lawsonia inermis - HENNA (Ar), ELAM, ILAM (T).

Breitenbach gives ELAM (T), HENUN (Sid), GHEDUR, ELLAN (Som), HENNA (Ar).

Cufodontis gives ELAM (Tigre), ALCANA, ALHENNA, HANNA, HENNA, HENNÉ, HENNEH, HINNA (Ar).

Dale and Greenway in Kenya Trees and Shrubs, 1961, note the name ELAM in the Boran language.

Note: The name ELAM, or one similar, is used also for *Impatiens tinctoria*, possibly for *Conyza spp.*, and possibly also for *Indigofera spp.*

Recipes using the root of ELAM have been placed under *I. tinctoria*. Strelcyn notes that Henna may be a (medicinal) equivalent to *I. tinctoria* (ENSOSILA) and from one text quotes: '... le médecin dit: si l'on (couvre) tout le corps avec ses feuilles chaudes (bouillies) comme on fait avec celles d'ənšosəlla, c'est bon'.

The name HINA (G), or one similar, is also used for *Indigofera spp.* which according to Cufodontis is used for colour. This name is never used for *Impatiens tinctoria*. Possibly the two recipes from Harrar refer to *Indigofera spp.*, though Arabic words are used in Harrar town and the true Henna is both readily available and in common use.

Recipes

(a) Dry the leaves of HENNA, grind them, and mix with water.
(b) Grind the leaves of HINA and boil them.
(c) Leaves of HINA are sliced by beating them, say, with an iron rod. Add water to them. After two or three days a reddish liquid is produced.
(d) The powdered dried leaves of HINA are mixed with water and some sugar is added. Apply to hands. Too much turns the hands red; if applied moderately the dye will give a red-pink colour.
(e) Add lemon juice to the crushed leaves of HINNA.

Sources: Ismael Omer (Hararghie) gives recipe (a). He thinks that HENNA is the Galla name. He describes the colour as 'pale' and used by Muslim girls.

Idris Osman Samra (Eritrea) also gives recipe (a) and the name HENNA, adding: 'The plants grow wild in hot climates. Collect the leaves and sun dry them. Used by women for colouring hands and legs. The colour is brown. Some men use HENNA to dye their beards. If HENNA is kept in water for more than three days, it will have a dark colour.' (This source also gives ILAM, describing it as exactly like HENNA and growing in 'cold climates'. Recorded under *Indigofera arrecta*.)

Abdulkadir Ahmed (Eritrea) also gives recipe (a) and the name HENNA (T). Like the above, he also describes ELAM (T) as another plant

Bissrat Fesahaye (Tigre) gives recipe (b) and the name HINA, which he describes as a small tree. The resultant dye, he says, is a fast black colour.

Million Abebe (Hararghie) gives recipe (c) and the name as HINA (G); he describes the colour as brown.

Ogbai Zeru and Hadgu Bariagaber give recipe (c), Ogbai Zeru making the distinction that HINA applies to leaves and ILAM to the roundish root the size of a fist. (Probably *Impatiens tinctoria*.)

W/t Azeb Desta (Harrar) gives recipe (d).

Ogba Michael Tekle giving recipe (e), says that the leaves are ground like flour and command a higher price than ILAM (see under *Indigofera arrecta*). He adds that their application is similar, except that 'some chemicals, probably lemon' are put into the HINNA. It is a lowland plant not very commonly used by the rural peoples (in Eritrea) and often sold by 'Somali peoples in Harrar province'.

Osyris abyssinica - CHIRET, KARET, KERET, GARAD (A), TAKAZALLEH (A, T), GERAR, KARATH, KERAZ, *KERETSE* (T), ASASO (G-H).

Breitenbach gives: DINGEROKOLA (A), WATO (G), GARAD, KERETZ, TOKASILLA (T), ASSOSSO (Som). East African Sandlewood.

Cufodontis gives: GERAR, KARATH, KERAS, KERATH, KERAZ, KERETZ, QÉRRAS, KHARAT, TOKAZALLÉ (A), CHERAZEGHI (Hamasen), WATO (G). He adds that the bark is used for tanning.

Note: Some of these recipes may apply to *Acacia spp.* Breitenbach gives *Acacia etbaica* as KARAT (T) and *Acacia nilotica* as KARAD (Arabic), adding: 'Bark and seed shells contain tannin extracts which are used for tannery purposes as well as against dysentery.' Mooney gives *Acacia etbaica* as KARATH (T) and *Acacia spp.* as KIRRAT, KRATT (Harrar). Cufodontis gives *Acacia etbaica* CARAT, GARATH, QARATH (Tigre in Mensa). Lemordant gives kärät as a 'petit arbuste aux fruits rouges' and adds that the leaves are used to perfume vessels (see recipe (d) of the Qottu Galla). He also gives qärät as *O. abyssinica*, specifying that it is used in tanning.

Historical notes: Lefebvre, T., in *Voyages en Abyssinie*, 1845, Vol. III, p. 241 on leather: 'Preparation des djendié. C'est un cuir rouge, teint avec l'écorce de l'arbre appelé "Kerate", ... Le jus de citron est le seul mordant employé.' P. 244 on dyers: '... dans la province du Sémiène, les cuirs en rouge avec l'écorce du Kerate'.

Cecchi, A., in *Da Zeila alla frontiere del Caffa*, 1886, Vol. I, p. 293: 'Per colorare le pelli in rosso, il "faki" si serve della scorza dell'albero chiamato "Kerate" ...'

Girard, A., in *Souvenir d'un voyage en Abyssinie*, 1873, p. 241-2: 'Les teinturiers abyssins teignent les cuirs et les lames, ainsi que le chanvre ... Ils font également usage de l'écorce du Kérate, ... pour teindre les cuirs en rouge.'

Von Heuglin, M. Th., in *Reise nach Abessinien*, 1874, p. 328: '*Osiris habessinica*? zum Rothfärben'.

Mérab, P., in *Impressions d 'Ethiopie*, 1929, Vol. III, p. 402: '... en teinturie ... les feuilles et l'écorce de "Kerat" (*Acacia ethbaica*) qui colorent en orange les étoffes, les peaux, les meubles, les calebasses "intus et extra"... L'acacia ethbaica est surtout utilisé par les musulmans.'

Recipes

(a) Collect some bark from the KERET tree. The bark should be cooked, and then left soaking for some days.
(b) Chop the leaves and boil KERETSE (T). Filter after three days. A fast colour for reddening leather *ankelba* (sleeping cradles) and *silicha* (grain containers).
(c) The bark of the stem is mixed with a small quantity of water for a few days. It produces a concentrated red ink which is used by tanners for dyeing leather.
(d) Chop and fry KERET leaves; then boil them and, together with HINNA (*Lawsonia inermis*), keep the dye in gourds for fifteen to twenty days to impregnate them with the colour. Afterwards smoke the gourds with olive wood to make them sweet for carrying water. Used by the Qottu Galla.
(e) Cut and grind the leaves of WATO (G). Add water and boil. This gives both red and yellow.

Sources: Bayelegne Meshesha (Bale) gives recipe (a).
Negussie Wolde Sellassie (Addis Alem) gives recipe (a) for red ink.
Mengesha Yibrah (Begemeder) gives recipe (b), the information about its uses and the name KERETSE (T).
Taddesse Woldemeskél gives recipe (b), adding: 'It can be used for fast dyeing of clothes.'
Taddesse Mengistu (Gojjam) gives recipe (c).
Wo Turuwork Teffera gives recipe (d).
Ismael Omer (Hararghie) gives recipe (e).
Assefa Bedada (Wollega) gives recipe (e): 'They bring these leaves to Addis Ababa and sell them to use for dyeing.'
Loul Ghebre Yohannes gives the name KERETSE (T) as 'used for colouring leather red'.
Memereh Kirkos Wolde Mariam (Gondar) says that QARAT, (using the explosive Q) is used to dye monks' garments, and that it slowly fades to yellow.

Pterolobium stellatum - KANTUFA, KOINTR (A), KUONTAFTA-TEH (T), KANTUFA (G-H), GARU (G-Sid).

Note: Breitenbach gives *Maerua aethiopica* as KONTR (A), and *Entada abyssinica* as KANTAFA (A).

Historical notes: Johnston, C., *Travels in Southern Abyssinia*, 1844, p. 167: '... covered with an ox skin tanned with the bark of kantuffa, which gives to this kind of leather a red colour.' P. 373: '... strewed the pounded bark of the kantuffa acacia (Pterolobium lacerans) after which cold water was added which in a few days became a strong red infusion.' P. 373: 'It is very probable that the celebrated Morocco leather, derives its bright red colour from the bark employed in tanning, being obtained either from the kantuffa or the Adal tree [*Acacia sp.*] for both of these trees give a very red colour to the skins that are prepared with their bark.'

Harris, W. C., *The Highlands of Aethiopia*, 1844, Vol. II, Appendix, p. 415: 'Pterolobium lacerans (*Kantuffa*) is an impenetrable hedge-shrub abounding in Efat: the bark gives a red dye for leather.'

Recipes

(a) Crush the leaves of KENTEFFE or KONTIR with stones. Boil with water and then filter.

(b) Crush, grind, and then boil KANTAFFA leaves in water to which a little oil or butter has been added, which helps to fix the colour.

(c) As in the recipe above, but with the addition of BERETTAR (iron slag or iron filings from the blacksmith).

For more complicated uses with other ingredients, see the recipes for Black Ink.

Sources: *Ato* Tekle Haile, Empress Menen Handicraft School Dye Shop, gives recipe (a), describing it as a good traditional red-brown colour, especially for leather.

Aleqa Wolde Medhin gives recipe (a) describing the colour as dark red or 'velvet'.

Wo Almaz Kebede gives recipe (b) saying that it is used to dye clothes for mourning and also for the dark colours in basketry.

Ato Imeshaw Temtin (Bulga, Shoa) gives recipe (c), adding that it is frequently used to decorate mule harnesses.

Rhamphicarpus spp. usually ***heuglinii*** — YESET LEB.

Identified by name alone by Mr M. Gilbert, the Herbarium, H.S.I.U. The literal meaning is 'a woman's heart'. Mr Gilbert says that this name is used for conspicuous but short-lived white or pale pink flowers, e.g. an *Ipomoea*, which is used medicinally.

See Red Ink recipe of *Marigeta* Tsege Tsefa Hunezu.

Rosa spp. - KEGGA, TESGYE-REDA (A).

Mooney gives *Rosa abyssinica* - KAJA, GHAGA, KEGA, TIGYE REDA, TSEGYE-REDA (A), GA-AGA (Harrar).

S. Gilbert gives *R. abyssinica* as KEGA (A), GORA, INKWOTO (G Shoa), GAKA (T). She describes the name TSIGEREDA (A and T, from Ge'ez and Ancient Greek) as being used for cultivated roses, including *R. richardii* which is often found in monastery and church gardens in Tigre and Ethiopia.

Note: Recipes (a) and (b) below appear to apply to cultivated roses.

Recipes
(a) Rub the petals on paper.
(b) Squeeze the juice of the petals into a bottle and mix with water.
(c) The pounded seeds of KEGGA give yellow.
(d) The bark of the roots of KEGGA is boiled for red-brown.

Sources: Recipe (a) given from Kaffa, Wollega, Sidamo, Illubabor and Bale.
Arega Gebre (Kaffa) gives recipe (b).
W/t Azeb Desta (Harrar) gives recipe (c).
Ato Imeshaw Temtin (Bulga, Shoa) gives recipe (d).

Rubia discolor - *ENCHIBIR, MINCHIRER, ?ANCHURURA* (A), SANKA, SCHANNOEH, SCHANKOKH, SCIANNOKH (Agau), SECHNEN, SCENENN, SSEEHHINIEN, SSEHHNEN, SSEKHENEN, ZECHNEN, ZEKHNEN (T), *?CHIMERI* (G), *BARTUTA* , *BARTUSHE* (Kambatigna)
With the exception of those in italics, names are taken from Cufodontis. See under Blue.

ENCHIBIR was identified from name alone by Dr Amare Getahun, Alamaya, H.S.I.U. CHIMERI (G) is a possibility, judging by the word's similarity to the Amharic; also by the recipes. BARTUTA and BARTUSHE (Kambatigna) were identified from specimens by Mr M. Gilbert, The Herbarium, H.S.I.U.

Lemordant gives ənç̣əbabər as the name of an unidentified plant, and mənç̣ərər, mənç̣ər, manğärurit, mənç̣ərurit as *Rubia discolor*, adding that it is a plant 'qui donne une poudre colorante rouge. La racine pilée et mise à bouillir dans l'eau et le beurre, sert à teindre divers objets rouge.' It is also used as a medicine for pain and coughs.

Historical notes: Von Heuglin, M. Th., in *Reise nach Abessinien*, p. 328: '"Seh'inien", eine Rubiacee, scheint nicht Kultiviert zu werden und dient zum Färben von Haaren'.

Mérab, P., in *Impressions d'Ethiopie*, Vol. III, p. 402: 'La garance (*Rubia tinctoria*) abonde dans le pays à l'état sauvage: on ne l'utilise pas comme source de couleur, ce qui se comprend, car son emploi est assez complexe.' (Cufodontis describes *R. tinctorum* as being once common but rarely grown today.)

Recipes

(a) The root of ENCHIBIR should be hammered with a stone, powdered, and mixed with boiling water. 'Used for red ink for writing.'

(b) As a wool dye for deep pink, using alum and salt, boil for two or three hours and leave to cool in the dye bath.

(c) INCHIBER or MINCHERER - 'they are the same' - together with *mashega* (animal skin glue) and sometimes scarlet earth, are boiled for ink and as a dye for clothes.

(d) Take the root of CHIMERI and wash it clean. Pound it wet; do not sun dry. Cook with a little water in an earthenware pot covered with a dung lid (as in making *injera*, local millet bread). Used for *mesob* (bread basket) colours. Fast.

(e) Grind CHIMERI 'properly' after drying in the sun, and then dissolve the powder in water. 'It is a very nice red colour. It will not run off in washing.'

Sources: Arega Gebre (Kaffa) gives recipe (a) and the name ENCHIBIR.

Sig. Gilardi and *Ato* Tekle Haile, Empress Menen Handicraft School Dye Shop, give recipe (b) and the name ENCHIBIR.

Wo Tschai Haile gives recipe (c), saying that the glue may be omitted when dyeing clothes.

Eshete Degefu gives both recipes (d) and (e), and the name CHIMERI (G), which has not yet been identified.

Geremew Getahun (Bale), during a conversation with the writer concerning names, gave ANCHURURA as being the same plant.

Ato Alemayehu Moges (Gondar and Addis Ababa) gives the name MINCHIRER (A) in his recipe for Red Ink.

An old man in the School of Fine Art, Addis Ababa, describes the colour as violet.

No recipes using butter, as described by Lemordant, have been collected.

Rubus spp. - ENJJORRI, INJORRE (A), GURA (G-W), *GORA* (G).

Recipe: Squeeze the juice from ripe juicy fruit.

Sources: Shiferaw Fufa (Wollega) gives both INJORRI (A) and GORA (G) and says the juice is used to dye cloth, baskets, chairs and other artifacts.

Tesema Ta'a (Wollega) states that some fruits give black and others red. 'Children use it to decorate their clothes.' Mulat Admassu says: 'Very good red colour ink by mixing with water.'

Ismael Omer (Hararghie) gives GORA (G).

Wondimagegn Tefera (Illubabor) gives both INJORY (A) and GORA (G).
Gebre Behute (Gojjam).

Rumex nervosus - AMBATSCHO, EMBACHO, IMBACHO (A), ASOT, HAHOT, HAKOT (T).

S. Edwards gives Imbwa*ch*o (A), *D*angego (G-Shoa), *H*a*h*ot (T).

See use of EMBEWACHO in Red Ink recipe of *Abba* Kidane Mariam.

Saccharum officinarum - Sugar Cane - SHINKUR-AGEDA, TIN-KISH (A), SHUNGKORA (A, G).

Recipe: Chewed sugar cane is placed in a metal container and a small fire lit under it. When smoke starts to rise from the sugar cane, women hold their hands over it and rub the smoke into the skin. Next morning the hands are red as if dyed with Henna, the colour lasting for about a month.

Sources: *Wo* Yewoinishet Alli (Harrar), who says this is practised by the women in Hararghie.
W/t Azeb Desta (Harrar).

Schefflera sp. - ADIM (T).

Mooney gives *Schefflera abyssinica* KETEMA, KUSTEA (A), HAR FATO (A, G), GEDDEM (T), AFATO, BUTO GATAMA, GATEMA (G), GETHEMA (G-W), MAFRATU (G-H), ORONKEH (G-Sid).
Schefflera volkensii GEDDEM (T), BUTO, GATAMA, GATEMA, LOUKEH (G), KOKORA (G-Lekemti).

Ato Alemayehu Mogus describes the ADIM tree as resembling a KOSSO tree (*Hagenia abyssinica*). Dr Tewolde Berhan, The Herbarium, H.S.I.U., suggested asking *Ato* Alemayehu Moges whether ADIM was another name for GEDDEM. He agreed that it was.

Recipe: The roots of ADIM (T) and of HEMBO-HEMBA (T) (*Cassia singueana*) trees are soaked to produce red.

Source: Yehdego Melles (Tigre).

Solanum nigrum - TIKUR-AWITT (A), ACHO (K).

S. Edwards in *Some Wild Flowering Plants of Ethiopia*, p. 34, explains that *S. nigrum* is known as Ne*ch* (white) awi*t* when unripe, and *K*eyee (red) or Ti*k*ur (black) awi*t* when the red or black berries are ripe and edible. It is often spoken of simply as AWIT or AUTT, as also is *Physalis peruviana*, the Cape gooseberry, which is sometimes given the adjective *ferenje*, or foreign.

Recipe: The red or purple berries of AWIT together with red soil are used to make red, probably for manuscript ink or as one of the ingredients in recipes for red inks and paints.

Source: *Aleqa* Desta Teklewold.

Ximenia americana - MELAHÓ, MELLAU, MELLUCH, MELLUK, MELLUOH, MELO (T), MALHEYFA, MELHETTA (Tigre), MULOO (Bogos), MUHNO (Dembesan), ANKOHÉ, ANKOI, UNGUAI, UNKWOI (A). Names taken from Cufodontis.

Recipes

(a) Cut out the red inner part of the bark of the tree known as MELOCH or MELO (T), which looks rather similar to an olive tree, and carefully grind the bark in a *mokecha*. Mix with animal urine 'and some other spices'. A fast colour used for dyeing *ankelba* (sleeping cradles).

(b) INKOI bark is boiled together with GUMERO (*Capparis sp.*) roots for a rose-coloured tint.

Sources: Aregawi Woldemariam (Eritrea) gives recipe (a) but cannot remember the other 'spices'.
Ato Imeshaw Temtin (Bulga, Shoa) gives recipe (b).

Unidentified - *AGOR GOBASH* (A).

Wo Almaz Kebede endeavoured to obtain a specimen for identification, but unfortunately was unsuccessful. She described it as a 'spring flower', appearing after the long rains, smaller than *Bidens spp.*, growing no taller than 50 cm. She described the flowers as 'vase shaped', pointing upwards and usually yellow, but thought they may also be red or dark violet.

See Red Ink recipe of *Memereh* Kirkos Wolde Mariam, who specified the use of yellow flowers.

Unidentified - *EJEAKELA, INCHAKELE* (A).
Described as having very short leaves.

Recipes

(a) Simply rub the article to be coloured with the leaves.

(b) Press the leaves to get red 'ink'.

Sources: Bezabeh Shibeshi (Sidamo) gives recipe (a) and INCHAKELE (A).
Masresha Haile gives recipe (b) and EJEAKELA.

Unidentified - *ETAIBETER, YETEBETERA* (Guraginia).
Described as a 'herb' found in Gurage country.

Recipes
(a) The leaves are cut and boiled for one or two hours. A colour resembling that of blood is obtained.
(b) A dye for cotton.

Sources: Dendir Dansamo (Sidamo and Shoa) gives recipe (a) and the name YETEBETERA.
Ato Tekle Haile, Empress Menen Handicraft School Dye Shop, gives recipe (b) and the name ETAIBETER. He describes it as a greenish dye for cotton, but this may be a slip of the tongue. (He certainly knew the plant as a dye stuff but may have made an error concerning its colour; he was discussing several other plants at the same time.)

?Pavonia sp. - *TOHGO* (G).

The leaf is said to be the same size as that of an adult eucalyptus. Mooney gives *Pavonia sp.* THOGO (K).

Recipes
(a) Cook the leaves with water.
(b) Rub the leaves on paper.

Sources: Yilma Kassa (Illubabor) gives recipe (a).
Endale Taddesse (Illubabor) gives recipe (b).

Acacia bussei - GALOL (Som-Harrar).

Historical note: Burton, R., in *First Footsteps in East Africa*, 1894, Vol. I, p. 144, writing of the Halimalah valley near Abasso: '... and a kind of acacia, here called Galol. Its bark dyes cloth a dull red, and the thorn issues from a bulb which, when young and soft, is eaten by the Somal, when old it becomes woody and hard as a nut'.

No recipes relating to its present use were collected.

Cordeauxia edulis - GUT (Som), the fruit YEHEB (Som).

Cufodontis gives: GUD, GUDA, GUDE, GUT (Som) and the fruit GHIEHEB, GIAEB, GIEHEB, YEHEB, IEEBB, IEE-EP, IEHEB, IIEB, IEEBB, YEHIB, etc. Shrub or small tree once widespread N-NE of the Webi Shebele in the Ogaden and Somalia.

The dye from this plant was used by the Italians as a substitute for Alizarin during the last war.

Source: Dr P. R. O. Bally, the Herbarium, Nairobi, who wrote an article in the Swiss journal *Candollea*, Vol. XXI, Geneva, 1966, to this effect. He also (verbally) described to the writer his work on this dye plant after the Italians left Ethiopia.

No recipes received in respect of its present use.

Discopodium penninervium - ALUMA, ALARARO, AMORAROO, HAMAMARO, HAMARARO, HAMORARU (A), AMBOROCHE (G), GUADA (T). Names taken from Cufodontis.

Historical notes: Harris, W. C., in *The Highlands of Aethiopia*, 1844, Vol. II, p. 411, states: 'Atropa arborea (*Amoraroo*), the red juice of whose berry is used by the Amhara women to stain their palms and nails, are common hedge-shrubs in Shoa.'

This appears to be a mistaken identification.

Savoia-Aosta, L. A. di, in *La esplorazione dello Uabi-Uebi Scebeli*, p. 432, on *Discopodium penninervium*: 'Nello Scioa è detta *Hamoraru o Hamararo*, e il succo rosso dei frutti à usato dai nativi per tingersi le mani e le unghie: in amarico *Alumà*, tigrignà *Guadà*.'

No recipes detailing its present use were collected except in the Black Ink recipe of *Aleqa* Desta Tekle Wold.

Unidentified - HADDIE.

Cufodontis gives: *Commiphora erythreaea*, va. *glabrescens* HADDI (Som), HABAGHADDI, HABBAK-HADDI (Som); *Commiphora spp. div.* HADDI (Som); *Maerua spp.* HADÉ (Danakil); *Cadia purpurea* KADI (Ar).

Breitenbach gives: *Hagenia abyssinica* KOSSO (A), DUSHA, FIETO, HETO (G), HABBI (T).

Historical note: Valentia, G., in *Voyages and Travels*, 1809, Vol. III, p. 162: 'Coarse carpets are made in Samen, and at Gondar, from the wool and hair of the sheep and goats, which are dyed red, and light blue; the former from a tree called Haddie...'

No recipes collected.

RECIPES FOR RED INKS

Red Ink recipe of *Ato* Alemayehu Moges of Gondar and Addis Ababa. Take:

the roots of MINCHIRER (*Rubia discolor*)
the roots of INSOSILLA (*Impatiens tinctoria*)
the bark of KERET (*Osyris abyssinica*)
the roots of GIRSHIRIT (possibly *Conyza sp.* or *Impatiens rothii*).

Chop and pound these ingredients, keeping all the juices. Pound three times a day, every day, alternating with sun-drying. An oily mixture should result. If time presses, *mucha* (gum, usually of *Acacia sp.*) may be added. Painters add red soil to this mixture. Red ink is preserved in dry form. When required for use, water is added and the mixture left for about three days, to 'leaven' it.

Red Ink recipe of *Memereh* Kirkos Wolde Mariam of Gondar. At Maskaram (after the rainy season, generally towards the end of September) gather:

TSIGEREDE ABEBE (*Rosa spp.*) flowers
ADAI ABEBE (*Bidens spp.*) flowers
AGOR GOBASH (unidentified) yellow flowers.

Dry the flowers and pound them to powder. Add pure water and *mucha* (gum) from GRAR (*Acacia sp.*). Continue pounding, adding more pure water. Sun-dry for six months. This is a red ink for manuscripts.

Red Ink recipe of *Abba* Wubu Kidane Mariam of the National Library. Translated by *W/t* Tsehai Berhane Selassie. Boil in lemon juice:

the roots of ENCHEBER (*Rubia discolor*)
the flowers of YEBAHER-TEF (*Eragrostis sp.*)
the flowers of EMBEWACHO (*Rumex nervosus*)
the flowers of QENTEFA (*Pterolobium stellatum*).

Add some butter and then filter.

Red Ink recipe of *Marigeta* Tsege Tesfa Hunezu of Debra Tabor, Gondar, calligrapher in H.I.M.'s Scriptorium at the Old Palace, Addis Ababa. Translated by *W/t* Tsehai Berhane Selassie. Take:

Red soil
MENCHIRER (*Rubia discolor*)
YESET LEB (*Rhamphicarpus*, usually *heuglinii*).

The three are mixed with water and *mucha* (gum, usually from *Acacia sp.*) added later. Put these in a wooden container and knead with cold water for three months in the sun. The *mucha* (gum) is dissolved in water in the sun. Filter it and add to the red ink. Earth should not be boiled or it will become darker in colour.

BROWN

Acacia etbaica	SERAW (T)
Beta vulgaris	KEYE-SIR (A)
? Brucea antidysenterica	TAMECHE (K)
Carissa edulis	AGAM (A)
Coffea arabica	BUNA (A)
Dombeya goetzenii	SHAWUKO, SHUKO (K)
? Lagenaria sinceraria	BURI (G)
Ocimum sp.	DAMAKESI (A)
Syzygium guineense	DOKHMA (A), LEHAM (T)
Iron slag or iron filings	BERETTAR (A)

Acacia etbaica - GRAR (A), KARATH (T), SUGSUG, YUBEH, YURAR (?).

Breitenbach gives: GRAHR (A), SERAU, ARAT, KARAT (T), SUG-SUG, YUBI (Som).

Recipe: The dried bark of SERAW (T) is soaked in water. After some days the water changes colour to khaki. SERAW is used for tanning and the red substance produced is called MAHAZEL (T).

Sources: Ogbai Zeru and Hadgu Bariagaber, who give the colour both as red and khaki.

Beta vulgaris (cult. Beet) - KEYE-SIR (A), *HUNDE DIMA* (G).

Recipe: Skin, chop, and put in water and boil. Produces a moderately fast colour.

Sources: various.

? Brucea antidysenterica - *TAMECHE* (K).

Breitenbach gives for *Brucea antidysenterica*: WAGHINUS (A), TAMICHA, ABABO, GUMANIO, DADATU (Arussi), MELITA (T), HADAWI (Som).

The bark, roots, fruits and leaves of this tree are used for medicinal purposes. The fruit is bright red.

Recipe: Take the fruit of the tree TAMECHE (K), pound it, and use it raw to obtain a brown colour.

Source: Bekele Wolde Gabriel (Kaffa): 'The people of Kaffa (Bonga) decorate different furniture with them.'

Carissa edulis – AGAM
and
Jasminum abyssinicum — TEMBELEL

Recipe: Collect the fruits of AGAM and TEMBELEL and squeeze them together for a brown-coloured liquid.

Source: Getachew Haile (Bale).
Seeds of *Jasminum* are used in the black ink recipe of *Abba* Wubu Kidane Mariam.

Coffea arabica, Coffee — BUNA (A, T, G), GIA (Sh).

Recipes

(a) Coffee leaves are boiled for about thirty minutes. The cotton is boiled with the leaves and can afterwards be woven for the *telat*, the coloured border to the *shamma*. A common practice in Hararghie.
(b) Coffee leaves are gently fried before being boiled as in recipe (a). This is a recipe of the Qottu Galla of Hararghie.

Sources: *Wo* Alli (Harrar) gives recipe (a).
Wo Turuwork Teffera gives recipe (b).
Marigeta Tsege Tesfa Hunezu uses coffee beans in his recipe for Black Ink.

Dombeya goetzenii - WOLKAFFA, WULKAFFA (A), ULKIFFA (G), DANESSA (G-K, G-Bale), SHAUKO (K).

Recipe: Take bark of the tree SHAWUKO (K), pound it and cook it with water for brown.

Source: Bekele Wolde Gabriel (Kaffa).

? Lagenaria sinceraria (*Cucurbitaceae*) - BURI (G), BOYE (Wollamo).

Tentative identification by name alone, of Dr Amare Getahun, Alamaya.

This plant was described by the sources as a small climbing plant with soft leaves.

Lemordant gives *buya* as an unidentified plant, and quotes Guidi as saying that it is a climbing plant which is edible.

Recipe: Cook the root of BURl (G) to obtain a reddish-brown.

Sources: Wondimagegn Tefera (Illubabor) gives the name BURl.
Eshetu Kebede (Illubabor) adds: 'the root is edible', and 'it stains the mouth when eating.'

Ocimum sp. - DAMAKESI (A), DAMAKASSIE (G).

Mooney gives *Ocimum suave*: DEMAKASI, YEMITSCH-MEDANIT (A), ABONATA, ABBU-NEDDIA, NEHRA (T), KUKO (Sh), and for *Lippia abyssinica*: KASI, DAMMA-KASSEH (A).

Cufodontis gives: *Ocimum lamiifolium*: DAMACHES, DAMAKHÉR (T), DAMACASÈ, DAMACSÈ, DARGU (G).

Lemordant gives: *Ocimum lamiifolium, O. menthaefolium*: DAMÄKASÉ.

This plant is used to relieve stomach cramps; also for eye infections, headache and even rabies.

Recipe: Take some DAMAKESI, a herb, and squeeze the leaves without cooking them, to obtain brown.

Sources: Temam Ibrahim (Illubabor), who gives DAMAKASSIE (G).
Yilma Kassa (Illubabor).
Endale Taddesse (Illubabor).

Syzygium guineense - DIGHITTAN (?A), DOGHMA, DOKHMA (A), LAHAM, LEHAM (T), BADESSA, BEDASSA, KURRUNFULI (G), GOFU, OCHU (G-W), WORARIKO (G-Sid), OKA (Wollamo), A-ACHA, DUANCHO (Sh), GASI?

Recipes

(a) The bark of DOKHMA can be used as a brown dye. Prepared as other bark dyes, with alum and salt.
(b) The bark of DOKHMA is boiled with water and used to dye clothes for mourning. Nothing is added to the water.
(c) Dig up the roots of DOKMA and boil with water for a yellow dye.

Sources: *Ato* Tekle Haile, Empress Menen Handicraft School Dye Shop, gives recipe (a).
Wo Tsehai Haile gives recipe (b).
Wo Turuwork Teffera gives recipe (c).

Iron slag or iron filings - BERETTAR (A).

Recipe: BERETTAR is obtained from a blacksmith. It must be ground to powder, and then put into a container with water. SINDEDO (*Pennisetum sp.*) is gathered in October. The seed heads are held over a flame and burned off. This makes the process of splitting the SINDEDO easier. The split SINDEDO is then put into the pot of water with powdered BERETTAR, and boiled for about twenty-four hours. The dyed SINDEDO is then dried in the sun. Before using to make a dark pattern on basket work, it must be soaked in cold water to make it pliable.

Source: *Wo* Askale Aregaw of Menz. The colour obtained was described as being dark: perhaps purplish-brown, brownish-black, or sometimes a dark grey.

The use of BERETTAR also mentioned by *Ato* Imeshaw Temtin, is described under *Pterolobium stellatum* in the section under Red.

BLUE

Delphinium spp.	GEDEL-ADMIQ (A)
Euclea schimperi	DEDEHO (A), KILLIAW (T)
Rothmania urcelliformis	DIBO (G, ?K)
Rubia discolor	ENCHIBER (A), BARTUTA (Kambat)
Trifolium acaule	WAJIMA (A)
Unidentified, *? Myrsine africana*	ARBA KECHEMO

Delphinium spp. - GEDEL-ADMIQ (A).

Name supplied by Dr Tewolde Berhan, the Herbarium, H.S.I.U.

D. dasycaulon - ZELLUM-DEBOSSOM (T).

Recipe: The flowers of *Delphinium wellbyi* and *D. dasycaulon* are gathered between September and November, dried, and mixed with the paste used for tattooing.

Source: Mrs Sue Edwards, I.A.R., who states that this is the custom in the north of the country.

Euclea schimperi - DEDEHO, MIECHA, MIESA (A), KELLAU, KILLIAW (T, ?A), GUM (T-Mensa), DUBOBIS (Harrar), DADAHO (G-H), KANKO (G-Sid), HANKU (Gugi).

see Green.

Note: The fruit is known as *ZEITO, ZEYATO* (T).

Recipes

(a) Squeeze the fruit and stir with warm water.
(b) Use the squeezed fruit juice with no water added.
(c) Take DEDEHO (A) and YEDEDEHA ABEBE (the flowers of DEDEHO), squeeze and add water. A fast colour.
(d) Take the flowers and fruit of DEDEHO and also add a little of the bark. Gives a dark bluish green.
(e) The seeds of DEDEHO mixed with the fruit of AGAM (*Carissa edulis*) will give a true blue.

Sources: Yacob Kassahun gives recipe (a), the name DEDEHO, the colour as violet, and says that it is fast.
Taddesse Abu (Tigre) gives recipe (b), the name DEDEHO and the colour as blue.
Tsegahun Mebrahtu (Tigre) gives recipe (c) mentioning no colour.
Aleqa Gebre Selassie (Selale) gives recipes (d) and (e).

Aleqa Wolde Medhin (Menz) gives recipes (d) and (e).
Teckle Tesfazghi (Eritrea) gives 'fruits of a tree ZIETO or KILAU (T). The colour is used for dyeing cloth and it shades slowly.'
Mengistab Tesfa Egzi (Tigre) gives 'a fruit of a tree called KELIAU (T). The fruit is known as ZEYATO. The colour is bluish.'
Gebre Yohannes (Tigre) gives KILIAW (T) but no colour.
Ogbai Zeru and Hadgu Bariagaber describe the liquid of ZETO as 'blackish'.
Woreda Isaac (Eritrea) gives the fruit of keli:au as blue.
W/t Azeb Desta (Harrar) gives the colour as black.

Rothmania urcelliformis - BARAQUÀ (probably A), DIBO (G), KUKO (Sid), OKEN (Chako).

Names given by Cufodontis.

Note: Dr Tewolde Berhan, the Herbarium, H.S.I.U., found this species to be fairly widespread in the south-west, and in Illubabor examined some plants on the northern slopes of the Qoncho river. The slash is orange-yellow; the fruit when cut is first cream, then yellow and finally changes to a bright orange. The stain left on the hands later turns blue. The villagers told Dr Tewolde that the fruits are crushed and boiled in water and that they used this as a blue dye, the colour obtained resembling that of blue jeans (indigo). The people in this region are Gallas, and they call the plant DIBO. A plant specimen of DIBO was identified at the Herbarium, H.S.I.U. as *Rothmania urcelliformis*.

Recipe: Squeeze the DIBO fruit and boil to get blue.

Source: Bekele Wolde Gabriel (Shoa and Kaffa). (The probability is that the name DIBO is also (K).)

Rubia discolor - *ENCHIBIR*, *MINCHIRER*, MONCIERER, *?ANCHURURA* (A), SANKA, SCHANNOEH, SCHANKOKH, SCIANNOKH (Agau), SECHNEN, SCENENN, SSEHHINIEN, SSEHHEN, SSEKHENEN, ZECHNEN, ZEKHNEN (T), *?CHIMERI* (G), *BARTUTA*, *BARTUSHE* (Kambatigna).

See Red for identification of names.

The following recipes are for the seeds of what was described (orally) as a 'fragile climbing plant with darkish blue-black seeds.' A sample later brought in was identified by Mr M. Gilbert, the Herbarium, H.S.I.U.

Recipes

(a) Take the seeds of BARTUTA (Kambatigna) and crush them when they are fresh. Do not add water. Bottle and use for green ink. Fairly fast, but not used for dyeing cloth.

(b) Write directly with the seed of BARTUSHE (Kambatigna). It is a blueish colour.

Sources: Desta Gebre Michael (Kambata, Shoa), gives recipe (a). Michael Abamo (Kambata, Shoa) gives recipe (b).

Trifolium acaule - WAJEMA, WAJJIMA (A), WODIMA (A-Gojjam), MAGAD MAGET (T).

Recipe: Press the flowers of WAJIMA.

Source: Worku Ambie (Gojjam) gives this as a 'blue colour'.

? Myrsine africana - ARBA KECHEMO.

Tentative identification by Dr Amare Getahun, Alamaya, H.S.I.U., from name alone.

Cufodontis gives *M. africana*: CACCIAMU, CATSCIAMO, COATSCIAMO, KACAMO, KACHAMO, KATCHAMO, KOHATTSCHAMA (A), GATSCHANA, GUJAMO, KACAMO (G), etc.

Recipe: The fruit when collected is mixed with a little water for a blue colour.

Source: Mulat Admassu, describing ARBA KECHEMO as a 'herb'.

VIOLET

Carissa edulis	AGAM (A)
Musa sapientum	MUS (A)
Pisum savitum	ATER (A)
Rhamnus staddo	T'ADDO (A), TZEDO (T)
Striga sp.	AKENCHIRA (?A)
Syzygium guineense	DOKHMA (A), LEHAM (T)
Unidentified	KIHÉ TSILMÉ

Carissa edulis - AGAM, HAGAM (A), AGAMSA (G).

S. Edwards gives AGAM (T).

Mooney and Cufodontis both give DIGHITTA (A-Gondar) but *Ato* Alemayehu Moges (Gondar and Addis Ababa) says this is not correct. Cufodontis mentions that this plant is used as a specific against snake bite, and Lemordant that it is used for toothache. He states that the mashed branches can be used as a tonic, a cough medicine, and an abortifacient.

The thorns are frequently used as an instrument for tattooing.

Recipes
(a) Squash the fruit and place it in a container, adding a little water.
(b) The fruits may be boiled to produce a blue dye for cloth.

Sources: Recipe (a) is provided by:

Assefa Asghedom (Kaffa), who gives the colour as purple, adding: 'It can be prepared within a short period. It is difficult to avoid the colour from the cloth' (i.e. it is moderately fast).
Tesfaye Fodone (Illubabor), who gives the colour as 'reddish'.
Yacob Kassahun, who gives the colour as violet.
Mengistab Teckle Egzi (Tigre), who gives the colour as 'blueish'.
Million Abebe (Hararghie), who gives the colour as purple.
Tsegahun Mebrahtu (Tigre), who gives the colour as 'dark pink'.
Taddesse Abu (Tigre), who gives the colour as blue.
Gebre Behute (Gojjam).
Alayu Gorfe (Harrar), who says: 'People press its seed and use it as ink.'
Gebre Yohannes (Tigre).
Arefa Gebresadik (Tigre).
Getachew Haile (Bale) says that mixing AGAM and TEMBELEL (*Jasminum abyssinicum*) will produce brown.
Aleqa Wolde Medhin (Menz) gives recipe (a) as making violet and recipe (b) blue. See *Euclea schimperi* for his recipe for true blue.

Musa sapientum, the Banana - *MUS*, MUZ (A).

Recipe: When cutting off branches, collect the juice, which gives a strong dark violet.

Sources: Altaye Gizaw, who thinks the colour is not fast, but adds: 'This gives you a very strong colour which does not even dissolve with petrol.'
Wo Yewoinishet Alli (Harrar).

Pisum sativum, Field Pea - ATTER, ATER, ATTUR (A), AINATER, TUKUR-ATER (T), ATARI-DANGULEH (G), ATERO, GISHI-SHATO (K).

Recipe: Squeeze the pea flowers for light violet.

Source: Yacob Kassahun.

Rhamnus staddo - *T'ADDO*, *T'EDDO* (A), *TZEDO*, *TZODO* (T).

Mooney gives THADDO (T).
Given by Lemordant as *Rhamnus deflersii*.

Historical note: Von Heuglin, M. Th., in *Reise nach Abessinien*, 1874, p. 329, mentions 'TADO' as *Rhamnus panciflorus*. See Green.

Recipe: The fruit of TSEDO (T) is used like ZEITO (the fruit of KILAU, *Euclea schimperi*). It is squeezed and used for dyeing cloth. It shades slowly.

Source: Tecle Tesfazghi (Eritrea).

Striga sp., probably ***hermonthica*** - *AKENCHIRA* (?A).

Identification by Mr M. G. Gilbert and Dr Tewolde Berhan, the Herbarium, H.S.I.U., from name alone.

Mooney gives MEZELLAM (T).

Described as a 'harmful weed to plants'.

Recipe: AKENCHIRA 'works as a very good colour. Its colour is purple'.

Source: Bezabeh Shibeshi (Sidamo).

Syzygium guineense - DIGHITTAN (?A), DOGHMA, DOKHMA (A), LAHAM, LEHAM (T), BADESSA, BEDASSA, *BEDESSA*, KURRUNFULI (G), GOFU, OCHU, OICHA (G-W), WORARIKO (G-Sid), OKA (Wollamo), A-ACHA, DUANCHO (Sb), GASI?.

Breitenbach gives: DOGMA, DIGHITA (A), BADESSA, BEDASSA (G), WORARICHO, DUANDO (Sid), AACHA, OCHA (Wollamo), LAHAM, ROHAZ (T).

Cufodontis gives DIGHITTAN (Gondar).

The tree is described as something like an olive tree, the fruits being the size of marbles. This fruit is often described as being the colour *woyntej* (red wine); purple, dark red or violet, and classically a deep rose.

Recipes

(a) The fruit should be squeezed and kept for some days before adding water. Can be used as 'ink' and is reputed fast.
(b) The juice of the fruit is mixed with soot.
(c) Boil DOKMA bark for a black dye for clothes.

See also under Black Ink.

Sources: Recipe (a) is given by:

Bissrát Fesahaye (Eritrea), using the name LAHM.
Altaye Gizaw, using the name DOKKMA, and describing the colour as blue-black. Fast colour.
Yilma Kassa (Illubabor), giving the colour as violet. Arefa Gebresadik (Tigre), using the name LIHAM.
Tesfaye Fodone (Illubabor), using the name DOKEMMA, and giving the colour as blue, and used as 'ink'.
Yacob Kassahun, giving the name LEHAM and the colour as violet.
Chane Addis (Eritrea), giving the colour as purple.
Mulat Admassu, giving the colour as blue: 'It is a very good ink.'
Wondimagehn Tefera (Illubabor) gives the name DOKMA (A) and BEDESSA (G), describing it as being a very 'helpful' colour and dye.
Gebre Yohannes (Tigre), using the name LIHAM (T).
Endale Taddesse (Illubabor) gives recipe (b) and says that DOKEMA alone gives the colour violet.
Wo Tsehai Haile gives recipe (c).

A type of long grass - *KIHÉ TSILMÉ* (T).

Described as the kind of long grass used for colouring and making baskets. According to Dr Tewolde Berhan, the Herbarium, H.S.I.U., the name refers to the dyed grass used for basketry. Ogba Michael Tekle

(Eritrea) says that baskets are made of a grass called ARGIHÉ (T), which is similar to *Eleusine sp.* AKERMA, AKRIMA (A, T, G), and called *KIHÉ TSELMA* (T) after decoration.

Recipe: When used to colour basket work KIHÉ TSILMÉ (T) is put damp on to the container and left for a week. The colour will by then have been absorbed and the article left to dry purple.

Source: Mengesha Yibra (Begemeder).

BLACK — Ingredients of BLACK INKS

Acacia spp.	GRAHR (A) See Black Inks and Gums and Resins
Allium sativum	NECH SHUNKURT (A) See Black Inks
Aloe sp.	ERET (A) See Black Inks
Arundinaria alpina	KARKAHA (A) Charcoal used in Black Inks
Arundo donax	SHIMBOKO (A) Charcoal used in Black Inks
Bersama abyssinica	AZAMER (A) See Black Inks
Boswellia papyrifera	ITAN (A) See Mediums and Black Inks
Calpurnia aurea	DEGETTA (A) See Black Inks
Capparis tomentosa	GUMERO (A) See Black Inks
?Clerodendron myricoides	MESERECH (A) See Black Inks
Coffea arabica	BUNA (A) See Black Inks
Cucurbita pepo	YEKURA HAREG (A), BUKO (K)
Datura stramonium	ASTANAGHIR, ATTEFARIS (A)
Discopodium penninervium	AMERAROO (A) See Black Inks
Dodonea viscosa	KITKITTA (A) See also Black Inks
Eleusine corocana	DAGUSA (A) See Black Inks
Eucalyptus spp.	BARZAF (A)
?Festuca sp.	GUASSA (A) See Black Inks
Guizotia abyssinica	NUG (A) See Black Inks and Mediums
Hordeum vulgare: Barley	GEBS (A) See also Black Inks
Jasminum abyssinicum	TEMBELEL (A) See also Black Inks
Juniperus procera	TEDH (A) See Gums and Black Inks
Lolium temulentum	INKERDAT (A) See also Black Inks
?Loranthus sp.	KURUD (T)
?Maytenus ovatus	GUARAVA (Gurag)
Nigella sativa	TIKUR AZMUD (A)
Olea africana	WAYRA (A) See Charcoal used in Black Inks and Mediums
?Ormocarpum muricatum or *?Ximenia americana*	ANQA ENCHIT (A) See Black Inks
Osyris abyssinica	KERET (A) See Black Inks

Pennisetum sp.	SINDEDO (A) See also Black Inks
Phytolacca dodecandra	ENDOD (A) See Black Inks
Ricinus communis	GOOLOH (A) See Black Inks
Solanum sp.	GABAR-EMBWAY (A) See Black Inks
Stephania abyssinica	ENGOCHIT (A) See also Mediums
Strychnos innocua	MARENZ (A) See Black Inks
Syzygium guineense	DOKHMA (A) See Black Inks
Trigonella foenum-graecem	ABESH (A) See Black Inks
Triticum aestivum: Wheat	SINDI (A) See also Black Inks
Verbascum sinaiticum	KATATENA (A) See Black Inks
Zehneria scabra	AREGRESA (A) See also Black Inks
Zizyphus spina-christi	KURKURA (A-Dire Dawa)
Unidentified	SHIFIT (T)
Miscellaneous	
Animal products: horn, ox-blood	
Black oxide of antimony	KUL (A)
Charcoal and soot	
BLACK INK RECIPES	

Allium sp. - NECH SHUNKURT (A).

Allium sativum, Garlic, is NECH SHUNKURT (A).

Mooney gives *Allium cepa* as being either KEYE (red) SHUNKURT, or NECH (white) SHUNGURT (A).

NECH SHUNKURT, sometimes added to recipes for paints, appears in the Black Ink recipe No. 1 of *Marigeta* Tsega Tesfa Hunezu.

Arundinaria alpine: Mountain bamboo - KARKAHA (A), LEMAN (G), CHINATO, SHIKARO, SHINATO (K).

Recipe: Crush the roasted stems and add water for a black colour.

For more complicated uses with other ingredients, see recipes for Black Inks.

Source: Gebre Igziabher Haile Mariam (Wollo) and others.

Arundo donax - MAKHA, SHAMBEKO (A), SHAMBAKO (G, K), SHAMBUKO (T).

Recipe: Roast the stems of SHIMBOKO. Grind and mix with water.

Sources: Various: also *Ato* Alemayehu Moges in his recipe for Black Ink.

Bersama abyssinica (Cufodontis) - ASAMER, ASAMR, ASAHMER (A), TELA (A-Kuonhi in N-H), ASCHAOM, BERSAMA, CORRSOMA, KORSEMMA, KORSOMMA, KURSSUMA (T), BOCHÓ (Maho), SOMBO (Yuka in N-H), ČIBIRU, EBERRAKO, KORAKA, LOLČIČA, LOLCISA (G), DOLKIISSA, JOUMEFOC (G-Kuonhi), TEBERACCO, TEBERAKO, TEBERACCO (Sidamo).

Used in the Black Ink recipe of *Abba* Wubu Kidane Mariam.

Capparis sp. - GUMERO (A).

Probably *C. tomentosa* - GIMERO, GUMARU, GUMURO (A), ANDAL, ANDEL, ANDELO (A, T), GOMBOR (Harrar).

See also under Red and Green.

Recipe: The branches of GUMERO (A) with their leaves, and the roots also, are roasted so that they become black like charcoal. This is an ingredient of the Black Ink of Menz.

Sources: *Aleqa* Gebre Selassie (Selale).
Aleqa Wolde Medhin (Menz).

? Clerodendron myricoides - MISSERITCH, *MESERECH* (A-Debra Sina), SURRUBATRI (T), AGHIO (K).

MESERECH is an ingredient in the Black Ink recipe No. 1 of *Abba* Wubu Kidane Mariam. This name may also apply to *Jasminum spp.*, but as Jasminum appears in the same recipe under the name of TEBELELE (probably a misspelling of TEMBELEL), *Clerodendron myricoides* seems a more likely identification.

Cucurbita pepo: the Pumpkin — BAHARKEL, DUBA, DUBBA, YEQUARA-ARAG (A), WUSHISH (T), BUKO, BUKEH (K).

Recipes
(a) YEKURA HAREG mixed with soot gives a good black.
(b) Pound and boil the root of BUKO (K) for yellow.

Sources: Masresha Haile gives recipe (a).
Bekele Wolde Gabriel (Shoa and Kaffa) gives recipe (b), but there may be a mistake in identification: in conversation he described BUKO as 'a tree'.

Datura stramonium - ASTANAGHIR, ASTENAGRT, ATAFARIS, *ATTEFARIS*, *ESSEFARIS* (A), *ASANGIRA*, *MENGY* (G), *ASTENAGIR*, *MESTANAGER*, *MEZERBOL* (T).

See Green for further information regarding names.

Recipes

(a) YE ASTENAGER FERE (Seeds of Astenager) 'will be fried on fire and crushed to make it powder. Then mix it with water to produce black ink for writing'.

(b) Pounded seeds of ATAFARIS give black and are used to polish shoes.

It is also used in some Black Ink recipes as a fly-repellant.

Sources: Arega Gebre (Kaffa) gives recipe (a).
W/t Azeb Desta (Harrar) gives recipe (b).

Discopodium penninervium - ALUMA, AMERAROO (A).

AMERAROO is an ingredient in the Black Ink recipe of *Aleqa* Desta Teckle Wold.

Dodonea viscosa - KITKITTA, KUTKUTA, TADACHU (A), TAHSSES, TASIS (T).

The roasted leaves of KITKITTA are used in the Black Ink recipes of *Memereh* Kirkos Wolde Mariam and *Liqua siltanet Abba* Hapte Mariam Workeneh.

KITKITTA and WAIRA (*Olea africana*), used together, make a charcoal for black ink. See the charcoal recipe of *Aleqa* Wolde Medhin; also his Black Ink recipe.

Eleusine corocana - DAGUSA (A), DAUCHO (K), BARANKIA (Boran).

DAGUSA is an ingredient of the *harur* or mixing liquid in the Black Ink recipe of *Liqua siltanet Abba* Hapte Mariam Workeneh. He states that it is better than the more common barley or wheat.

Eucalyptus globulus - BAHIR-ZAF, BARZAF (A, T), AKA-CHITTA, ATAKILI (G), *KALAMETUS* (T).

Recipes

(a) Take the bark of large fully-grown BARZAF trees. Dry and chop it and then boil fast for two or three hours, with added alum and salt (30 grams of each to 1 litre of water). Allow to cool. Makes a fast blue-black dye for wool.

(b) Take the leaves of BAHIRSAF, when pounded and mixed with water, gives a blue colour.

Sources: Recipe (a) is provided by Sig. Gilardi and *Ato* Tekle Haile of Empress Menen Handicraft School Dye Shop, who say that they used this recipe with success after the war, when no imported colours were available. Sig. Gilardi used to dye his trousers with it. *W/t* Azeb Desta gives recipe (b).

?Festuca sp. - *GEWASA, GUASA* (A).
or possibly *Hyparrhenia sp.*

An ingredient in the Black Ink recipes of Mekuria Degu, of *Abba* Wubu Kidane Mariam in his recipe No. 1 and of *Memhere* Hailu Man Wasinot, who describes it as a grass of the *dega*, or temperate regions, used for rope-making.

Guizotia abyssinica - NUG (A).

The oil can be used as a painting medium.

NUG is used with roasted barley, in a recipe given under *Hordeum vulgare*.

The 'fruits' are used in the Black Ink recipe of *Liqua siltanet Abba* Hapte Mariam Workeneh.

Hordeum vulgare: Barley - GEBS, GUBS, TEMEJ (A), GARBU, GEBRE (G), SHEKO (K).

Recipes

(a) Roast the grain, powder it, and mix it with NUG oil (*Guizotia abyssinica*). Leave the mixture in open sunshine for some days. Later take a little and dissolve it in water.

(b) Put damp barley mixture in a cloth and squeeze it.

(c) Roast unground barley and afterwards boil with water until it becomes a red 'tea'.

For more complicated uses with other ingredients, see recipes for Black Inks.

Sources: Wobetu Tadege (Begemeder) gives recipe (a).
Mengistab Tesfa Egzi (Tigre) gives recipe (a).
Arega Gebre (Kaffa) gives recipe (b).
Zena Yitbarek (Bale) gives recipe (c).
Other sources with unspecified recipes come from Illubabor, Gojjam, Wollo, and Begemeder.

Jasminum abyssinicum - MESSERIK, TEMBELAL, TERERAK, TIMBILAL, WEMBELEL, ZOHUN-KACHAMO (A), ADDI-ZELLIM, HABBI-SELIM (A, T), ABITEREK, HABBE-TEREK, HABBE ZELLIM (T), MESSIRICH (Harrar), ELKEMME, TEO (G), TEMBELEL, BILU (G-H).

Recipes

(a) Dry the leaves of TAMBELAL. Grind to powder and add clean water. 'The colour will be black and fine'. Fast.

(b) Dry the leaves and seeds (?fruit). Grind to powder and add some water. 'A very powerful black colour can be obtained which cannot be washed away.'

An ingredient of some recipes for Black Ink.

Sources: Eshetu Degefu gives recipe (a).
Ashagrie Tasew (Bale) gives recipe (b).

Lolium temulentum - ANDERDAD, ENKERDAD, INKERDAT (A).
Rye Grass, Darnel Grass.

Derived from the Ge'ez *kerdat*, meaning tares, or unwanted grains cast aside during threshing.

Sometimes used in the preparation of *talla*, (barley) beer.

Historical note: Harris, W. C., *The Highlands of Aethiopia*, 1844, Vol. II, Appendix, p. 403: '... grasses found in every pasture ... Lolium temulentum (*Enkerdad*), much dreaded as poison ...'

Recipes

(a) Take roasted grain of INKERDAT and powder it: leave the mixture for ten days in the sun. Mix with water and use for ink.

(b) As above, but including GRAHR *mucha*, gum of *Acacia sp.*

(c) Roasted INKERDAD together with the roasted stalks of SINDEDO (*Pennisetum sp.*) and SINDI (*Triticum sp.*) are ground together.

For more complicated uses with other ingredients, see recipes for Black Inks.

Sources: Very numerous. This appears to be the best-known ingredient of manuscript black ink, a considerable number of laymen believing it to be the sole ingredient. Not only the grain but the stalks are occasionally roasted for ink.
Yedehgo Melles (Tigre) gives recipe (b).
Aleqa Gebre Selassie gives recipe (c) as a painting colour.
Aleqa Wolde Medhin gives recipe (c) as a painting colour.

? Loranthus sp. - *KURUD, KIRID* (T).

Described as a 'parasite growing in hot climates in the west of Eritrea'.

Cufodontis gives *Loranthi sp.* - DAQALLA, DOQALLA, DEKALA (T), KERUD (Tigre), and all *Viscum spp.* as DAQALLA (T).

Alternatively it could be *Striga sp.* or *Osyris sp.*

Recipes
(a) The leaves are gathered and dried in the sun. They are then ground and boiled with water. The resultant dye is used to darken the palm tree leaves used for making baskets, huts, etc. Leave the palm leaves to darken for a week.
(b) Soak in water for two or three days. Used to blacken DOM (*Hyphaene ethbaica*) palm mats. Fast colour.

Sources: Idris Osman Samra (Eritrea) gives recipe (a).
Saleh Mahammed Mahmoud (Eritrea) gives recipe (a).
Abdulkadir Ahmed (Eritrea) gives recipe (b).

?Maytenus sp. - GUARAVA (Guraginia).

Breitenbach gives *Maytenus ovatus var. argutus* - ATAT, KAMU, KURAVA, TELALO, TALO (A), HATCHAT (T), KOMBOLCHA (G), DEGMUT, DUBODEIS (Som).

Mooney gives *M. ovatus* - ATATT (A), ADAD, ADDAD (A, T, G), HATCHAT, KAMO, KURAVA, THELALO (T), KOMBOLCHA (A, G), HOMBOLCHA (A-Ar), ANGITO (T).

Cufodontis gives *M. ovatus* - ADAD, ATAT (T, G), HATSCAT (T-Agau); *M. pubescens* - ADAD, etc. (T, A), KAMO, COURAVA, KURAWA, THELALO (T).

Recipe: The bark of this big tree is crushed, mixed with damp earth and buried beside a river. Any colour of earth seems to be satisfactory. It remains buried for three or four days in a damp hole. Gives a fine fast black, and is a good carpet dye.

Source: *Ato* Tekle Haile, Empress Menen Handicraft School Dye Shop.

Nigella sativa - ASMUD, AVOSEDDA, AZMUD-TIKHUR, KOR-AZMUD, TIKUR-AZMUD, TUKUT-AZMUD (A), AWOSETTA (T), GURRA (G-W), A-AFO (K).

It is possible that recipe (b) may apply also to *Nigella sativa*, though the name given is TIKUR KEMAM, literally 'black spice'. Some say that this name refers to the seeds of *Schinus molle*, the Pepper Tree, YE KUNDOBERBERE (A).

Recipes

(a) Take the spice TIKUR AZMUD, chop and grind it to powder. Mix with water and then filter. A fast colour used for dyeing cloth.

(b) Roast and grind TIKUR KEMAM and mix the 'flour' with some water to get black ink.

Sources: Taddesse Woldemeskel gives recipe (a).
Worku Ambie (Gojjam) gives recipe (b).

Olea africana - WEIRA, WOIRA (A), VIGREH, VOGHERA, WOGREH (T), EDIRSA, EGERSA (G), MEKICHU (G-Sid), WALAGA (?).

There are several other *Olea spp.* in Ethiopia.

The wood of *Olea spp.* is used both to make charcoal for drawing and to mix in Black Ink.

The fruit may be crushed and the oil used as a painting medium; also to prepare canvasses.

Mentioned in the Black Ink recipes: No. 2 of *Ato* Alemayehu Moges, No. 2 of *Abba* Wubu Kidane Mariam and No. 2 of *Marigeta* Tsega Tesfa Hunezu.

Pennisetum sp. - SINDEDO (A).

Recipe: The stalks of SINDEDO together with the stalks of SINDI (*Triticum sp.*) and INKERDAD (*Lolium temulentum*), are roasted and ground together for black.

See also recipes for Black Inks.

Sources: *Aleqa* Gebre Selassie (Selale).
Aleqa Wolde Medhin (Menz).

? Ricinus communis - HABLALIT, TSCHIAKMAL (A), GULLEH (T), GUM (T-Mensa), KELLA (T-Tigre).

GOOLOH leaves are used as a fly-repellant in the Black Ink recipe of *Aleqa* Desta Teckle Wold.

Stephania abyssinica - AREGAIT, ENGOCHIT, UNGOCHIT, YETAREGH, YETGURRU (A), ITSE YESUS 'Plant of Jesus' (Ge'ez), IAT-GURRA (A-Gondar), IDI-ATOUTA (G), KALALA (G-Bale).

Recipes

(a) Roast the flowers, and the root (which is the main item) as an ingredient for Black Ink.

(b) The ENGOCHIT flowers with their seeds also provide an oil which, when mixed with GRAHR *mucha* (gum from *Acacia sp.*) and TEDH *mucha* (gum from *Juniperus procera*), 'cooked' together and the filtered through a rag, produces a good painting medium.

Source: *Aleqa* Gebre Selassie (Selale) provides both recipes.

Strychnos innocua - ENGAUCHIA, MARENJ, MARIZ, MERENZ (A), HUNGUAKH-HEVEI (T).

MARENZ is an ingredient in the Black Ink recipe No. 1 of *Abba* Wubu Kidane Mariam.

Trigonella foenum-graecem - ABISH, AVISH (A), ABACHAM, ABACHEH (T), SHUNKO, SUNKO (G), GRARO (K).

ABESH seeds are used in the Black Ink recipe No. 1 of *Marigeta* Tsega Tesfa Hunezu.

(There is a possibility that ABESHO was intended, in which case this could be *Datura stramonium*, which appears in some medicinal or magical recipes under this name.)

Triticum aestivum: Wheat — SINDI (A), ADJAH (A, G), KAMADO (G).

Recipes: as for *Hordeum vulgare* (barley). Recipes include the use of both grain and stalks.
Simple recipes give roasted SINDI mixed with water and baked SINDI mixed with soot. More complicated recipes with other ingredients are to be found under Black Inks.

Sources: Numerous, but not from the southern Provinces.

Verbascum sinaiticum - DABAKADET, KATATENA (A), AMBO KANA (G-Bale).

QETETENA is an ingredient in the Black Ink recipe No. 1 of *Abba* Wubu Kidane Mariam.

Zehneria scabra - AREG RESA, *HAREG RESA* (A), HAFAFELO, HAFFAFALA, HAFFA-FALU, HASFAFALA (T), and possibly also *ITSE SABEK* from the Ge'ez *eṣ̂ä sabéq*. *HIDDA FITA* (G-Shoa), *UMBAO, IMBAU* (Wollega and Illubabor).

The above local names are from Cufodontis, with the exception of those in italics.

Negaso Gidade (Wollega) gives UMBAO and IMBAU as G. Wollega and Illubabor.

Adane Feyyisa (Wollega) gives HIDDA FITI as G. Shoa.

Lemordant adds that students drink the juice to obtain '... science et eloquence'. The plant is extensively used for medicinal purposes.

See under Green.

Recipes

(a) The leaves of AREGRESA are roasted and ground and the resulting powder mixed with a little water.

(b) Crushed AREGRESA leaves are mixed with roasted GEBS (*Hordeum vulgare*). Leave for twenty-four hours and then strain for a good black ink.

(c) The leaves are taken from the climbing stem. They are mashed so that they give juices. The juice is collected in a beaker and left exposed to the sun. Later the liquid becomes a greasy red substance. It is used as an ink for writing.

(d) The leaves and bark of ITSE SABIK are roasted and ground. After being dried and powdered they should be used with a gum medium for 'dark colours' for painting.

Sources: Worku Ambie (Gojjam) gives recipe (a).
Haile Kebede (Gojjam) gives recipe (b).
Taddesse Mengistu (Gojjam) gives recipe (c).
Aleqa Wolde Medhin (Menz) gives recipe (d).
Mentioned in the recipe for magic Black Ink supplied by M. Jacques Mercier.

Zizyphus spina-christi - NABK (A), ABATTARYEH (A), GABA, GABBA, GEBA, GHEBBA (A, T), GHEVA, GOBBA, KOSLET, KOSSILEH, KULKULL, LESARA (T), KUSSA (T-Acrur), KOSSILA (T-Mensa), KURKURA (A-Wollo, G-Ar). Known as KURKURA in Dire Dawa. Positive identification of specimen by *Ato* Getachew Aweke and Mr M. G. Gilbert, University Herbarium.

Recipes

(a) The leaves of KURKURA are pounded and the juice is collected. After washing the hair, rinse with this juice to make hair black.

(b) The leaves are first dried and powdered. One teaspoonful is used as a shampoo: it gives an excellent lather. Reputed to wash, condition and tint the hair.

Sources: *W/t* Azeb Desta (Harrar) gives recipe (a).
Wo Yewoinishet Alli (Harrar) gives recipe (b).

Unidentified - *SHIFIT, SHIFT.*
A tree.

Recipe: Pluck the tender leaves of the tree. Crush them on a smooth stone, sprinkling with a little water. Collect the thick liquid in a container. Add powdered charcoal or paraffin soot. Leave in the sun for a week. A good black ink.

Source: Tsegahun Mebratu (Tigre).

ANIMAL PRODUCTS

Horn - KEND (A).

Recipe: Horn of animals, blackened and powdered.

Sources: Mengesha Yibrah (Begemeder).
Tsegahun Mebrahtu (Tigre).
Sheep's horn is included in the Black Ink recipe of *Liqua siltanet Abba* Hapte Mariam Workeneh, and horn of a black sheep is one of the ingredients for ink for use on magic scrolls provided by M. Jacques Mercier.

Hair - Hair of black goat is an ingredient of the magic ink mentioned above.

Blood - Dried ox blood is mentioned in Recipe No. 3 of *Abba* Wuba Kidane Marian.

Ox blood is also used to decorate hides in the Jimma area.

Bones - Bones, especially bulls' bones, are used for brown.

Recipe: Burn until dark and then grind to a fine powder, which should be brown in colour. Mix with oil - one of the oils described under Mediums - and use for shading when painting.

Source: *Aleqa* Wolde Medhin of Menz.

BLACK OXIDE OF ANTIMONY

Kohl - KOOL, KUL (A), KUHLI (T).

Imported as a salve and as a cosmetic.

Historical note: Dr Mérab, in *Impressions d'Ethiopie*, 1929, Vol. III, p. 389, notes: '... en réalité, ce n'est pas de l'antimoine, mais un sulfure de plomb.'

KUL is available in most markets, as also is a black powder known as KULL (to rhyme with gull, mull), used to dye clothes black for mourning. A type of laundry blue, known by the name of LEEN and also used to dye mourning clothes, is occasionally called KULL.

Recipe: Crush, powder, and mix with water.

Sources: Numerous, referring mostly to KUL but at times to KULL also.

CHARCOAL

Charcoal - KESSEL (A), CHILATI (G).

Charcoal forms an integral part of nearly all the recipes for Black Ink. The principal sources for this charcoal are:

Arundinaria alpina, the Mountain bamboo - KERKAHA (A);
Arundo donax - SHIMBOKO (A), SHAMBUKO (T);
Olea africana, and possibly another *Olea sp.* - WAYRA, WOIRA (A).

Branches of other trees are also used. The following recipe of *Aleqa* Wolde Medhin of Menz, which includes both *Olea africana* and *Dodonea viscosa* - KITKITTA (A) - describes the preparation of charcoal:

> 'Dig a small pit and fill it with WAIRA and KITKITTA. Set fire to them, but as soon as they start to burn, cover the pit with leaves, and then earth. Press the earth down well so that no air can escape.'

Some inks are made almost entirely from charcoal. Elizabeth Wolde Mussie reports: 'When I visited one of the churches on one of the islands in Lake Tana, I asked how they prepared the colours for the paintings. They told me that all the colours except the black were imported. They said that they prepared the black colour from powdered charcoal, by boiling it with water, leaving it aside for some days, and then again boiling it. This process is repeated for six months: the colour is then ready for use.' *Note:* It is probable that the water used in this recipe was in fact a liquid made from boiling some type of roasted grain. This liquid is known as the *harur*, and is used as the mixing agent in most ink recipes.

Simple recipes

(a) Powdered wood charcoal mixed with water and oil.
(b) ENDOD (*Phytolacca dodecandra*) and kessel mixed together. (Powdered charcoal rubbed with the leaves of ENDOD is also used for preparing blackboards. See under Green.)

Sources: Very numerous, both for complicated and simple recipes.

SOOT

Soot - TELAT (A), TEKER (T).

Soot taken from the underside of a cooking vessel, usually from the almost flat *mitad*, is known as TELASHIT, TELASHET (A).

Grass soot from above the kitchen fire is TIKERSHA, TEKERSHA (A).

Soot forms an integral part of nearly all recipes for Black Inks. (Soot is also used for tattooing.)

Simple recipes

(a) Soot is mixed to a paste with water. Keep it for some days, and then add an equal amount of oil.
(b) Soot is mixed with wood charcoal, and well ground with a stone. Mix with water and keep in a bottle.
(c) Soot mixed with lemon and water.
(d) Soot mixed with butter.

Sources: Very numerous from all the highland areas.

RECIPES FOR BLACK INK

Black Ink recipe of *Memereh* Wolde Kirkos of Gondar, who demonstrated calligraphy on parchment at the Addis Ababa Book Exhibition, 1972. Take:

(a) KERKEHA kessel (charcoal made from *Arundinaria alpina*)
(b) *telat* (soot) from burning *gaz* (paraffin) under a metal vessel
(c) roasted SINDI (*Triticum aestivum*). (It was intimated that INKERDAT (*Lolium temulentum*) was really better for this purpose but less easy to gather.)

To the roasted SINDI (*Triticum aestivum*) add water so that it becomes slightly fermented. The liquid produced by the mixture is somewhat oily, and only the liquid is used with the two other ingredients, which are pounded together with a metal pestle and mortar. Place in sunlight.

This mixture is stirred and then sun-dried - water having been added as necessary - from between two to three months up to six months. It should look oily when ready.

Dry lumps are preserved and pounded with added water when required for use.

Black Ink recipe of *Memereh* Kirkos Wolde Mariam of Gojjam. Take:

(a) *telat* (soot) from burning *gaz* (paraffin) under a metal vessel
(b) roasted leaves of KITKITA (*Dodonea viscosa*)
(c) roasted outer skin of KERKEHA (*Arundinaria alpina*).

The three ingredients should be mixed together, reduced to powder and then put in a pot and boiled with water like porridge. Remove, and put into a container with GEBS (*Hordeum vulgare*) or SINDI (*Triticum aestivum*) *harur* (which is the liquid in which the roasted grain has been fermenting).

Alternately sun-dry and pound for five or six months.

The ink is better without any *mucha* (gum, usually from *Acacia sp.*) or *mashega* (glue made from boiling cow hide) added.

Black Ink recipe taken from *Tintawi ye ethiopia temert sineserat* (System of Ancient Ethiopian Education) by *Liqua siltanet Abba* Hapte Mariam Workeneh of Gondar. Translated by *W/t* Tsehai Berhane Selassie. 'How to make Black Ink' (pp. 267-8):

1 *Soot and Charcoal*
(a) Gather soot from various utensils such as *mitad* [flattish cooking vessels], bowls and the like.
(b) Burn the leaves of KETKETA [*Dodonea viscosa*], powder them and mix with the soot.

(c) Burn the leaves of QARAT [*Osyris abyssinica*], powder the burned leaves and mix with the soot.
(d) Burn the bark of QARQAHA [*Arundinaria alpina*], and mix with the soot.
(e) Burn the leaves of QANTAFA [*Pterolobium stellatum*], and mix with the soot.
(f) Burn the horn of a sheep, powder it and mix with the soot.
(g) Roast the fruits [seeds] of NUG [*Guizotia abyssinica*] until they become dark, and boil until it simmers, gather the vapour and knead, adding the vapour it. [?]
(h) Gather the gum of GERAR [*Acacia sp.*], and add for mixing well.
(i) Put all this into a clean container, powder very finely, and boil all of them together. Keep in a strong container and knead from time to time. It becomes dry with the passing of time and the kneading. Finally add the *harur*, and knead again when it is a bit softer.

2 *The harur or qitran* (tar) [the liquid used for mixing inks]:
harur is the grain chaff that gives lustre to the colour. Anything such as barley [*Hordeum vulgare*], wheat [*Triticum aestivum*] or ENDERDAD [*Lolium temulentum*] will do. The best is that of DAGUSA [*Eleusine corocana*]. Roast and boil it. A not very thin water of this [the water in which the chaff has been boiled] will be added from time to time. A hot temperature is necessary for the fermentation.

3 *Time*
The condition of kneading determines the fermentation. For its readiness, at least three months are necessary. Six months is the right time to have an ink which will not fade. Then it is pounded and kept for future use.

Red and green are made from varied soil and leaves in a similar way. Other colours are achieved by different mixtures of these two colours. Such colours never fade.

4 *Final stage*
The ink is kept in tablets or cakes, is finally made thinner in a horn container. It should be allowed to stand at least two days before use.

Black Ink recipe from *Tentawi yaqolo tamari* (The Ancient poor Student), by *Abba* Kidane Maryam Getahun. Translated by *W/t* Tsehai Berhane Selassie. Excerpt from p. 37:

'... although the ancient student had learning as his main duty, he had many handiworks he did in his spare time. Among these were (a) he helped himself by extracting ink by mixing the *harur* of wheat (*Triticum aestivum*), the soot from earthen and metal *mitad* [flattish cooking vessels], QANTAFA [*Pterolobium stellatum*] and *mucha* [gum, usually from *Acacia sp.*] ...'

Black Ink recipe No. 1 of *Ato* Alemayehu Moges of Gondar and Addis Ababa. Mix together:

telashit (soot, usually taken from the underside of a metal mitad, a flattish cooking vessel)
quantities of toasted DOKEMA leaves (*Syzygium guineense*)
toasted SHIMBOKO (*Arundo donax*)
toasted GEBS (*Hordeum vulgare*),
adding a little water.

After three or four days the mixture should become slightly fermented. Mix these ingredients together and add good *mucha* (gum, usually from *Acacia sp.*), which should look like beeswax. It is even better to use *mashega* (a glue made by boiling cow hide until it liquifies). Pound the mixture, sun-dry, pound and sun-dry, adding water, preferably that in which a roasted grain has been boiled, whenever necessary. Continue for six months, depending on altitude and climate. (The initial mixture may be prepared at a high altitude and then sent down to a lower altitude to be stirred and sun-dried, in order to hasten the process.)

Note on *mucha*: It is kept in tablet form, and generally it is exposed to sunlight for two or three hours, together with a little water, to soften it before use.

Black Ink recipe No. 2 of *Ato* Alemayehu Moges of Gondar and Addis Ababa. Take:

green twigs and shoots from WOIRA (*Olea africana*).

Fry them hard, or set light to them and then cover them and let the fire die slowly. Pound the resulting charcoal and add:

mucha (gum, usually from *Acacia sp.*)
telashit (soot, usually taken from the underside of a metal *mitad*, a flattish cooking vessel).

(N.B.: This is all that is written in the recipe, but verbally he added that *harur*, water in which roasted grain has been boiled, should also be added.)

Proceed as in recipe (1).

This ink is generally quicker to make than that in recipe (1) but may take up to six months.

Black Ink recipe No. 3 of *Ato* Alemayehu Moges of Gondar and Addis Ababa. Take:

tekersha (grass soot from above the cooking fire in the thatched-roof kitchen), and add
quantities of mitad *telashit* (soot from the underside of a *mitad*, a flattish cooking vessel). Mix with
harur, water in which roasted grain has been boiled.

This is the quickest recipe to make.

Black Ink recipe of Mekuria Degu of Shoa. Roast:

KNTEFFA (*Pterolobium stellatum*)

on a thin metal plate, and then grind it to flour with cast iron. Add:

SINDI (*Triticum aestivum*) flour
GUASSA (*Festuca sp.*) flour, and
ESSEFARIS (*Datura stramonium*) fruit, crushed.

All these are mixed with water and stirred daily for a month. After this long process, the water will have evaporated. Keep the mixture dry, and mix with water in small quantities as required.

Black Ink recipe of *Aleqa* Desta Teckle Wold, as told to *Ato* Daniel Tewafe and *Ato* Gebre Tsadik Wolde Meskel, of the National Museum, Addis Ababa.

The *harur* (mixing fluid) for the ink is prepared with:

one bottle of GEBS (*Hordeum vulgare*)
half a bottle of SINDI (*Triticum aestivum*)
half a bottle, or a little less, of INKERDAT (*Lolium temulentum*).

These should be roasted like medium-roast coffee, then a little water added.

A *mitad* (flattish cooking vessel) is first brushed with a feather to remove all the dust and the *telashit* (soot) then collected from the underside.

Added to the *telashit* (soot) are:

the leaves of KITKITTA (*Dodonea viscosa*) and
AMERAROO (*Discopodium penninervium*), both roasted,

possibly with some

GOOLOH (*Ricinus communis*) leaves, which keep away flies from the ink, and/or
ETSE FARIS (*Datura stramonium*), which possesses a similar property as a fly-repellant.

The leaves, soot and *harur* are mixed 'like bread'. Some GRAR *mucha* (gum from *Acacia sp.*) may be added.

Black Ink recipe No. 1 of *Memhere* Hailu Man Wasinot of Selale, Sire Medhane Alem. Calligrapher in H.I.M.'s Scriptorium, the Old Palace, Addis Ababa. Translated by *W/t* Tsehai Berhane Selassie. Take:

> GUASSA (probably *Festuca sp.*), [described as] grass of the *dega* or temperate regions, which is used for rope-making.

Cut about five handfuls of this grass. Burn it and take the ash to the stone mortar. Grind it well in the deep stone (the mortar), with another small stone (the pestle). Now mix well with water. Add the water little by little. After it is well mixed with water, fry:

> SINDI (*Triticum aestivum*)

to a dark colour. Add water and boil the fried SINDI. This is called YESINDI *harur*. This SINDI *harur* should be added little by little to the mixture for a whole month. By then the ink will be ready.

The test for ink - to find out if it has been well prepared and is ready - is whether or not it stains the stone. If it does not, more SINDI *harur* should be added every four days, well mixed. When it is ready, take a spoonful at a time and shape it and dry it on a tin sheet. The ink should be mixed with SINDI *harur* when it is to be used.

Black Ink recipe No. 2 of *Memhere* Hailu Man Wasinot of Selale, Sire Medhane Alem. Calligrapher in H.I.M.'s Scriptorium, the Old Palace, Addis Ababa. Translated by *W/t* Tsehai Berhane Selassie. Take:

> young SINDI (*Triticum aestivum*) plants.

Cut the young plant into small pieces and burn it on a *berat mitad* (metal *mitad*, a flattish cooking vessel), or some other thing, until it is dark. Grind it well in the mortar. Then mix it well by adding water little by little. Do this for eight days. After eight days, fry:

> GEBS (*Hordeum vulgare*)

until it is dark, and boil it in water while the GEBS is still hot. Mix with this *harur* little by little every hour. When it is ready it will stain the stone. Then take it out from the mortar little by little, using a spoon, and dry on a tin sheet.

Black Ink recipe No. 3 of *Memhere* Hailu Man Wasinot of Selale, Sire Medhane Alem. Calligrapher in H.I.M.'s Scriptorium, the Old Palace, Addis Ababa. Translated by *W/t* Tsehai Berhane Selassie. Take:

> ANKWA (probably *Ormocarpum muricatum* or *Ximenia americana*)
> AREG RESA (*Zehneria scabra*)

GABER INDOD (*Phytolacca dodecandra*) seed
KENTEFA (*Pterolobium stellatum*) leaves.

Split the wood into smaller pieces, and before it is too dry, burn it into charcoal. Pound it well. Boil together the other plants, and start mixing the charcoal powder with the mixture. After some days, take the leaves of KITKETA (*Dodonea viscosa*) and GEBS (*Hordeum vulgare*), and fry them dark. Mix them with water and add to the other mixture.

If the mixing has been continuous, the ink will be ready in twenty days. Dusty places should be avoided.

The test to see if it is ready is its texture, and the way it looks when you allow it to flow. After it is ready, take it little by little with a spoon, and dry it on a sheet of tin.

Note: All types of writing ink are prepared in the sun. If a stone mortar is hard to find, a wide vessel of earthenware may be used.

Black Ink recipe No. 1 of *Marigeta* Tsege Tesfa Hunezu of Debra Tabor, Gondar. Calligrapher in H.I.M.'s Scriptorium, the Old Palace, Addis Ababa. Translated by *W/t* Tsehai Berhane Selassie.

(a) Mix a *tasa* (small tin) of soot from a metal *mitad* (flattish cooking vessel) with water, and let it stand for eight days in a wooden bowl.
(b) Stir with a WAIRA (*Olea africana*) stick for one month.
(c) After a month or two, add YEGEBS *harur* (liquid made from roasted GEBS (*Hordeum vulgare*) fermented in water); roast the GEBS until it is black, add water while [the GEBS] is still hot; boil, strain, and add to the *telat* (soot) in the wooden container.
(d) Keep in the sun and stir it for three months, in the following manner:
- at first, for ten minutes every hour;
- then for ten minutes every two hours, and so on until the mixture requires stirring only for ten minutes every twelve hours.
(e) It is kneaded with the GEBS *harur* added to it. The mixture is kneaded until the ink is extracted (until ready?).
(f) When the ink is ready, it is marked by its lustre, and by the fact that it becomes sticky on the stirring stick.
(h) (sic) Then roast GEBS [to make a new] *harur*, until it becomes dark. Roast the following until they are 'over-roasted':

GRAR *mucha* (gum from *Acacia sp.*)
a handful of BUNA (*Coffea arabica*)
ABESH seeds (*Trigonella foenum-graecem*) and
NECH SHUNKURT (*Allium sativum*).

Boil these with the GEBS *harur*. When cool, filter through a piece of cloth and add to the ink that is being kneaded. When the ink is 'extracted', boil the GEBS *harur* with roasted NECH SHUNKURT

(*Allium sativum*) and ABESH (*Trigonella foenum-graecem*) and filter. Boil the ingredients mentioned above once more: boil in the GEBS *harur*, adding two fruits of GABAR EMBEWAY (*Solanum sp.*, possibly *S. marginatum*) cut in two halves. The EMBEWAY is added to prevent flies from entering the ink while one writes.

(g) (sic) Then, when it is thin enough, boil the ink. When it is cool, filter and keep in the sun. (The residue rests below.)

(i) The resultant mixture is left in the sun as well as in the cold. It becomes thick. Dry that in the sun and keep in a container. If the ink should lack lustre when it is thinned or when one writes with it, add filtered GEBS *harur*.

The soot from *gaz* (paraffin) is prepared in the same manner.

Black Ink recipe No. 2 of *Marigeta* Tsege Tesfa Hunezu of Debra Tabor, Gondar. Calligrapher in H.I.M.'s Scriptorium, the Old Palace, Addis Ababa. Translated by *W/t* Tsehai Berhane Selassie.

Break a large quantity of thick and reddened WAIRA (*Olea africana*). Fill a huge (earthenware) pot with it, and cover the mouth of the pot with soot from *gaz* (paraffin) and *mucha* (gum, probably from *Acacia sp.*).

Turn upside down and burn the pot (i.e. light a fire round it). Keep a smaller pot underneath. The steam will settle in the smaller pot. Take this out (i.e. the small pot) and write with the ink from it.

Black Ink recipe No. 1 of *Abba* Wubu Kidane Mariam of the National Library. Translated by *W/t* Tsehai Berhane Selassie.

(a) Roast the seeds of TEBELELE (probably a misspelling of TEMBELEL - *Jasminum abyssinicum*) and the grass known as GEWASA (probably *Festuca sp.*);

(b) Roast:

KETEKETA (*Dodonea viscosa*)
MESERECH (*?Clerodendron myricoides*, given by Mooney as MISSERITCH (A-Debra Sina))
AZAMER (*Bersama abyssinica*)
ERET (*Aloe sp.*)
QETETENA (*Verbascum sinaiticum*, given by Mooney as KATATENA (A))
MARENZ (*Strychnos innocua*, given by Mooney as MARENJ, MARIZ, METENZ (A)).

(c) For mixing [i.e. to make the *harur* liquid], roast ENKERDAD (*Lolium temulentum*) and NACH SINDE (*Triticum sp.*, white wheat) and let them stand in water.

(d) Pound and grind the plants mentioned above [mix with the liquid derived from (c)], and knead all together.

Black Ink recipe No. 2 of *Abba* Wubu Kidane Mariam of the National Library. Translated by *W/t* Tsehai Berhane Selassie.

Extraction of colour: Ingredients:

Yimitad *telat* (soot from under a *mitad*, a flattish cooking vessel)
leaves of QANTAFA (*Pterolobium stellatum*)
leaves of KETKETA (*Dodonea viscosa*)
leaves of WAYRA (*Olea africana*)
leaves of DEGETTA (*Calpurnia aurea*)
leaves of TED, the tender one (*Juniperus procera*).

Method: Dry the ingredients in the sun and then roast them on *Yeberat mitad* (a metal *mitad*) until well browned. Add the *mucha* (gum) of GRAR (*Acacia sp.*).

Boil these all together. Place the residue [resultant mixture] on broken earthenware. Stand it in the sun, kneading from time to time. When quite dry add YEGABES *harur* [liquid made from roasted GEBS (*Hordeum vulgare*) fermented in water] and knead occasionally for three to six months.

Black Ink recipe No. 3 of *Abba* Wubu Kidane Mariam of the National Library. Translated by *W/t* Tsehai Berhane Selassie.

(a) Roast and pound
YEGERAR *mucha* (gum from *Acacia sp.*) and
ANQA ENCHAT (probably wood of *Ormocarpum muricatum* or *Ximenia americana*).

(b) Boil
the leaves of KETEKETA (*Dodonea viscosa*)
the leaves of QANTAFA (*Pterolobium stellatum*)
the leaves of AREGERESA (*Zehneria scabra*)
the seeds of GABAR EMBEWAY (*Solanum sp.*, possibly *S. marginatum*, given by Mooney as GEBER-ENBUYEH (T))
the leaves of ENDOD (*Phytolacca dodecandra*).

Use this liquid to mix the kasal (charcoal, burnt product), item (a).

or

Dry the blood of an ox on YEBERAT METAD (a metal *mitad*, a flattish cooking vessel) by leaving it in the sun for three days, or by burning. Mix with the water used to boil the leaves, as mentioned in (b) above, and knead.

Black Ink recipe of *Aleqa* Gebre Selassie of Selale and *Aleqa* Wolde Medhin of Menz. Take:

the roasted or charred roots and branches of GUMERO (*Capparis tomentosa*)
roasted SINDI (*Triticum sp.*)
roasted SINDEDO (*Pennisetum sp.*).

Pound them together with *mitad telashit* (soot from a *mitad*, a flattish cooking vessel).

Add:

boiled *mucha* (gum) from GRAR (*Acacia sp.*) or TEDH (*Juniperus procera*).

Grind between two mill stones. Allow the mixture to ferment in the sun. Every week the mixture should be kneaded, and a little more *mucha* added. Eventually it begins to smell and become sticky, turning a dark colour.

Black Ink recipe for use on magic scrolls in Tigre, supplied by Monsieur Jacques Mercier. Take:

telashit (soot) from a metal *mitad* (flattish cooking vessel)
rust of metal
horn of black sheep
hair of black goat
leaves of KITKITTA (*Dodonea viscosa*).

Roast and grind these ingredients. Add:

zebeb (a decoction of dried grapes in water, sometimes used in perfumes).

Leave for one month and then test.

Mix with *harur* of SINDI (*Triticum aestivum*) and add leaves of ITSA ABEK (*Zehneria scabra*).

Mix and sun-dry; also leave in the cold.

Add salt and butter.

RECIPES FOR COLOURED EARTHS

Earths of all colours are used in country districts for colour-washing houses, for dyeing clothing (especially for mourning), to serve as blackboard 'chalks' and, in some areas, to decorate gravestones. A variety of local names is used to describe these earths, but most are commonly referred to by colour alone. The recipes are very simple.

RED EARTHS

Red earths are dissolved in water for two or three days and well stirred. They are used for colour-washing houses and sometimes for dyeing clothes. Oil may be added to make the colour more permanent.

A red earth called *sariyan* (A) is mixed with paraffin 'for writing purposes', and very fine well-sieved red soil sometimes mixed in a bottle with boiling water to make 'school ink'.

In some areas there a firm, moist, sticky clays, known as *dewol* or *dewel* (A), which are preferred.

Galla people use a clay of this type called *dororo* (G) to paint on gravestones, claiming that it does not run off, even during heavy rains.

The pinkish red soil from termite colonies is also used.

YELLOW EARTHS

Recipes are similar to those for the red earths.

There is also an orange coloured 'soft rock', to which oil is added. It is used for painting houses and for dyeing clothes.

WHITE EARTHS AND CHALK

Note: Ethiopian chalk is generally diatomite.

To make blackboard chalk, pieces are soaked in water for some time and then taken out and allowed to dry before use. This is a common practice in the north of the country.

There is, in addition, a 'white soil that looks like salt, found under big stones', known as *bolebol* or *bole* (G). Sometimes the colour is yellowish or grey. This is also used as blackboard 'chalk', but without preliminary soaking.

For colour-washing houses, white earths are treated in the same way as the red earths.

One rather curious recipe for a greyish soft rock found near Harrar, mentions that 'when mixed with red soil it will give yellow. If more red soil is added, it gives green. If you add to this last some black soil, you will get a good black, or a dark brown if less is added'.

BLACK EARTHS

Historical note: Pearce, N., in *Life and Adventures in Abyssinia*, Vol. I, p. 196, describing how cloth is dyed for mourning: '...the cloth ... is again buried in a black mud, common in all plains, called *walkar*; after remaining buried three days, it is taken out and washed but still remains black'.

Black earths, which may be very hard, almost rock-like, are used to dye clothes for mourning. After the earth has been dissolved it is boiled and oil added to the boiling water, sometimes together with salt.

Some black earths are said to produce fast colours, but the powdered soft stone known as *dado* (G), used as a brown dye for clothes and also for hair, is not fast.

A mud called *tibru* (G), which is dug from beside river beds, is put into a pit and covered with wet leaves and left for a week, after which it becomes a brownish black. Said to be fast, it is used to dye reeds, wooden articles and cloth.

GREEN EARTHS

Many pockets of green earth are to be found. It is dissolved in large containers, as in the other recipes, and used for colour-washing. In some areas this soil is known as *temenahe* and is said to produce a fast light green 'comparable to any building paint in Addis Ababa'.

A mud produced from decaying vegetation known as *longi* (T) is found by digging beside rivers and lakes. It gives a blue-green, and is also used for colour-washing.

BLUE EARTHS

Pockets of blue earth are not easily found but they do exist, and in a range of blues from turquoise to deep ultramarine. Such earths are used more for painting than for colour-washing.

A recipe of *Marigeta* Tsege Tesfa Hunezu describes how to make 'blue ink', but this is used to colour the decorative *haräg* and for painting, not for calligraphy. This 'ink' is made from finely-ground blue earth, not clay and not stone. Any blue flowers may be added, failing which the earth alone may be used. Acacia gum is dissolved in water in the sun, filtered and added to the ground earth and water. It was stated that earth mixtures should not be boiled; boiling makes the colour darken.

PAINTING MEDIUMS

A wide variety of painting mediums may be used; some are gummy and some oily. They are employed both for the earth colours and the vegetable colours and also for such animal products as charred horn or bone. An artist will keep them all separately as blocks or in bottles or horns, according to the substance.

The following recipes for mediums, and the list of further mediums, were provided by *Aleqa* Gebre Selassie (Selale) and *Aleqa* Wolde Medhin (Menz).

(a) Mix together the *mucha* (gum) of GRAHR (*Acacia sp.*), the *mucha* (gum) of TEDH (*Juniperus procera*), and *itan*, resin of *Boswellia papyrifera*. Add a little water, and boil the three together. Filter through a cloth and when cold, add yolk of egg. Never add white of egg.

(b) Almost the same as the above: GRAHR *mucha*, TEDH *mucha* and ITAN *waha* (Incense water). The latter is kept in a bottle and mixed in when needed.

(c) GRAHR *mucha*, TEDH *mucha* and the seeds and flowers of ENGOCHIT (*Stephania abyssinica*), from which an 'oil' can be extracted. Boil all these together, then filter through a cloth.

(d) RET, especially SETE RET (*Aloe spp.*), is cooked 'like a cabbage' and then pounded to extract a sort of greasy jelly. Filter. This is not a first-class medium, but gives a good shine.

(e) YETIT FERE WATET. Milk of cotton seed (*Gossypium sp.*).

(f) KINCHIB (*Euphorbia sp.*, probably *schimperi*).

(g) KULKWAL (*Euphorbia sp.*, probably *candelabrum* or *abyssinica*).

(h) SUF (probably *Helianthus annuus*, though *Carthamus tinctorius* is also known by the same name).

(i) NUG (*Guizotia abyssinica*).

(j) SELIT (*Sesamum indicum*).

(k) YE WOIRA zayt. Oil of *Olea africana*, the African olive tree. This is also used for oil painting.

(l) TELBA (*Linum usitatissimum*, Linseed). Used for oil painting.

(m) The collected essence from steaming green wood.

DISTILLED MEDIUMS

Recipe of Asres Yanesaw given in his book *Taqami meker* (Useful Advice), in the chapter on 'The use of plants', p. 25. Translated by *W/t* Tsehai Berhane Selassie.

> 'Chop wood and put it into a cracked pot. Close the top of the pot with a piece of earthenware cut to size and turn upside down and fit to the mouth of an unbroken container that is underneath. Take

care that the cracked container will not break due to heat. Stick the two containers together with mud, adding a lot of it so that the steam cannot escape. Put much wet ash on the upper part over the mud. Then make a fire with a lot of wood. When the cracked container gets very hot, the steam from the freshly cut wood in it will drip into the container beneath it. Mix this with powdered flowers of different colours or with coloured earths, powdered and ground. Let it ferment. When added to *qeteran* (tar), the sooty ingredients of ink according to balance, it produces colour that shines like glass.'

Aleqa Wolde Medhin describes another method of extracting wood or plant essences:

'A pot filled with water is placed on the fire. Over its mouth is a basket dish on which is piled the pieces of wood or other sub stances. Covering the basket dish and its contents is a dome-shaped piece of earthenware, from one side of which a bamboo pipe protrudes. The essence is caught in a bottle affixed to the end of the bamboo pipe.'

GUMS AND RESINS

Gum - MUCHA (A), TSERRI, ENDIDA (T), HARPPEH (G), from:

Juniperus procera - TEDH, TICH, SADD, ZADD (A), NERRET, TUYA, ZEDDI (T), GATIRA (G), INESA (G-Bale), YUDES (Boran), SIGIB, TAJIB (Som).

from various *Acacia spp.* - GRAHR (A),

and from *Boswellia papyrifera* — YETAN-ZAF (frequently simply ITAN) (A), ANQUAH (Tigre), KAFAL, GALGALAM (Ar).

Notes by Breitenbach:
Acacia senegal is the present-day gum arabic tree.
Acacia nilotica var. tomentosa was used in former times as the source of gum arabic.
Acacia sieberiana yields a clear gum of good quality.
Boswellia papyrifera is subject to tapping, whereby the aromatic resin yielded is used as incense. Two other *spp.* also yield aromatic resin, exuded from the buds.

These gums, forming an essential part of most ink recipes, are also used in the preparation of both vegetable colour paints and earth colour paints. They may also be used in dye-baths. It is said that GRAHR *mucha* is found in large pieces and TEDH *mucha* in small pieces.

Good *mucha*, after preparation, should look like beeswax and is kept in tablet form. When required for use, it is exposed to sunlight for two to three hours, together with a little water to soften it.

Historical notes: Harris, W. C., in *The Highlands of Aethiopia*, p.180, describing a scribe: 'His ink is a mucilage of gum-arabic mixed with lamp black. It acquires the consistency of that used in printing, and retains its intense colour for ages.'

Johnston, C., in *Travels in Southern Abyssinia*, p. 291: 'The ink is a composition of powdered charcoal and gum arabic or myrrh, with a little water.'

Burton, R., in *First Footsteps in East Africa'*, writes of *Acacia seyal*: '... the tall "Wadi" affords a gum useful to cloth-dyers.'

ANIMAL SKIN GLUE

Animal skin glue, used in the preparation of canvasses and sometimes to mix with colours, is known as MASHEGA. The following recipe for its preparation is provided by *Aleqa* Gebre Selassie (Selale) and *Aleqa* Wolde Medhin (Menz):

> 'Take a fresh skin. Turn it inside out and bury it in the ground for three days. On digging it up, take off all the hair and cut it into suitable pieces for boiling. Boil with water. Alternatively the skin may be boiled with the hair still on, and then the still-boiling mixture is poured through a cloth (which acts as a filter) into another vessel.
>
> If it becomes necessary to add more water, the water added must be boiling. The vessel is kept at a medium heat and must be stirred continuously to prevent sticking ['... all through the night ...']. Test to see if the glue is ready by lifting the stirring stick out of the pot. When a 'string' of glue hangs from the stick, take the pot off the heat. Keep stirring all the time it takes the mixture to cool, or it will stick to the pot. When quite cold take it out of the pot in thin strips and stretch it on boards to dry. It looks rather like liver when it is ready to be taken out of the pot.'

VARNISHES

Some of the painting mediums can also be used as varnishes. *Memereh* Kirkos Wolde Mariam of Gojjam uses *mashega* (animal skin glue) as a picture varnish. *Ato* Alemayehu Moges condemns this practice; he contends that the *mashega* may liquify and spoil the painting.

Aleqa Gebre Selassie and *Aleqa* Wolde Medhin say that a little egg yolk, if not added to the painting medium, may be used as a varnish to give a shine 'like oil paints have'. If too much is used it will flake off. A little wax is sometimes rubbed over a painting also to give a bloom or shine to it and to act as a preservative. Boiled *Aloe spp.*, filtered, are also used.

Bibliography

Abbadie, Arnaud d', *Douze ans dans la Haute Ethiopie*, Paris, 1868.
Almeida, M. de, *Some records of Ethiopia 1580—1645*, Beckingham, G. F. and Huntingford, G. W. B., (Hakluyt Society), New York, 1961.
Asres Yanesaw, *Taqami meker* (Useful advice), Addis Ababa, 1958, E.C.
Baeteman, J., *Dictionnaire amarigna-francais*, Dire-Daoua, 1929.
Beke, C. T., *Diary of a journey in Abyssinia, 1840—1843*, British Museum, ADD 250 A and B, 251, 252, 254.
Bent, J. T., *The Sacred City of the Ethiopians*, New York, 1896.
Breitenbach, F. von, *The Indigenous Trees of Ethiopia*, 2nd edition, Addis Ababa, 1963.
Bruce, Sir J., *Travels to Discover the Source of the Nile*, Edinburgh, 1790.
Burger, W. C., *Families of Flowering Plants in Ethiopia*, Stillwater, Oklahoma, 1967.
Burton, Sir R. F., *First Footsteps in East Africa*, London, 1894.
Castanhoso, M. de, *The Portuguese Expedition to Abyssinia 1541—3*, translated by R. S. Whiteway, (Hakluyt Society, second series, No. 10), London, 1902.
Cecchi, A., *Da Zeila alla frontiere del Caffa*, Rome, 1886-7.
Chojnacki, S., 'Some notes on the history of the Ethiopian flag', *Journal of Ethiopian Studies*, Vol. I, No. 2, Addis Ababa, 1962.
— 'Notes on art in Ethiopia in the 15th and early 16th century. *Journal of Ethiopian Studies*, Vol. VIII, No. 2, Addis Ababa, 1970.
— 'Notes on art in the 16th century', *Journal of Ethiopian Studies*, Vol. IX, No. 2, Addis Ababa, 1971.
Cohen, M., *Documents Ethnographiques d'Abyssinie*, Le Puy-en Velay, 1920.
Conti Rossini, C., 'Un codice illustrato eritreo del secolo XV', Africa *Italiana*, 1, Rome, 1927.
Cufodontis, G., 'Enumeratio Plantarum Aethiopiae: Spermatophyta', *Bulletin du Jardin Botanique National de Belgique, Supplément 22-24*, Bruxelles, 1953-73.
Dale, I. R., and Greenway, P. J., *Kenya Trees and Shrubs*, Glasgow, 1961.
Dästa Täcklä-Wäld, *Addis yamarəñña mäzgäbä qalat*, Addis Ababa, 1970.
Edwards, S., *Some wild flowering plants of Ethiopia: an introduction*, Addis Ababa, 1976.
Gerster, G., *Churches in Rock*, London and New York, 1970.

Graziosi, P., *Le pitture Ethiopiche de Museo Nazionale d'antropologia et etnologia de Firenze*, Rome, 1932.

Girard, A., *Souvenirs d'un voyage en Abyssinie*, Cairo, 1873.

Griaule, M., *Le Livre de Recettes d'un Dabtara Abyssinin*, Paris, 1930.

Hapte Mariam Workeneh, *Liqua siltanet Abba, Tintawi ye ethiopia temert sineserat* (System of ancient Ethiopian education), Addis Ababa, 1962, E.C.

Harris, W. C., *The Highlands of Aethiopia*, London, 1844.

Heuglin, M. Th. von, *Reise nach Abessinien (1861, 1862)*, Gera, 1874.

Hotten, J. C., *Abyssinia and its People*, London, 1868.

Hunter, F. M., and Fullerton, J. D., *Reports on Somali Land and the Harrar Province*, Aden, 1885.

Jelenc, D. A., *Mineral Occurrences of Ethiopia*, Addis Ababa, 1966.

Johnston, C., *Travels in Southern Abyssinia*, London, 1884.

Keller, C., *Über Maler und Malerei in Abessinien*, Zurich, 1904.

Kidane Maryam Getahun, *Abba, Tentawi yaqolo tamari* (The Ancient Poor Student), Addis Ababa, 1954, E.G.

Krapf, J. L., *Travels, researches, and missionary labours in Eastern Africa*, London, 1860.

Lefebvre, T., and others, *Voyage en Abyssinie*, Paris, 1845-51.

Lemordant, D., *Les plantes éthiopiennes*, Addis Ababa, 1960-61.

McKenzie, J. C., *Art Teaching for Primary Schools in Africa*, Nairobi, 1966.

Mérab, P., *Impressions d'Ethiopie*, Paris, 1927-29.

Montandon, G., *Au pays Ghimerra 1909—1911*, Paris, 1913.

Mooney, H. F., *A Glossary of Ethiopian Plant Names*, Dublin, 1963.

Pankhurst, R., *An Economic History of Ethiopia*, Addis Ababa, 1968.

Parkyns, M., *Life in Abyssinia*, London, 1853.

Pearce, N., *Life and Adventures in Abyssinia*, London, 1831, reprinted by Sasor, London, 1980.

Plowden, W., *Travels in Abyssinia*, London, 1868.

Poncet, C. J., *A voyage to Aethiopia made in the year 1698, 1699, and 1700*, (Hakluyt Society), London, 1949.

Powell-Cotton, P. H. G., *A sporting trip through Abyssinia*, London, 1902.

Purseglove, J. W., *Tropical Crops: Monocotyledons 2*, London, 1972.

Religiöse Kunst Äthiopiens: Religious Art of Ethiopia, (Institut für Auslandbeziehungen), Stuttgart, 1973.

Salt, H., *A voyage to Abyssinia*, London, 1814.

Savoia-Aosta, L. A. di, *La esplorazione dello Uabi-Uebi Scebeli, Dalle sue sorgenti nella Etiopia meridionale alla Somalia italiana [1928-29]*, Milan, 1932.

Schoff, W. H. (translator and editor), *The Periplus of the Erythraean Sea*, New York, 1912.

Spencer, D., 'In search of St Luke Ikons in Ethiopia', *Journal of Ethiopian Studies*, Vol. X, No. 2, Addis Ababa, 1972.

Stern, H. A., *Wanderings among the Falashas in Abyssinia*, London, 1862.

Strelcyn, S., *Medicine et Plantes d'Ethiopie*, Warsaw, 1968.

Tellez, B., *Travels of the Jesuits in Ethiopia*, London, 1710.

Valentia, G., *Voyages and Travels*, London, 1809.

Verdcourt, B., and Trump, E. C., *Common Poisonous Plants of East Africa*, London, 1969.

Winstanley, W., *A visit to Abyssinia*, London, 1881.

Wylde, A. B., *Modern Abyssinia*, London, 1901.

Yusuf Ahmed, 'An enquiry into some aspects of the economy of Harar, 1825—75', *Ethnological Society Bulletin* (1960), No. 10, Addis Ababa.

Index

www.ingramcontent.com/pod-product-compliance
Lightning Source LLC
LaVergne TN
LVHW050536100826
845148LV00002B/576